CAMBRIDGE COUNTY GEOGRAPHIES

ENGLAND

General Editor: F. H. H. Guillemard, M.A., M.D.

LEICESTERSHIRE

Cambridge County Geographies

LEICESTERSHIRE

by

G. N. PINGRIFF, B.A., B.Sc.

With Maps, Diagrams, and Illustrations

CAMBRIDGE
AT THE UNIVERSITY PRESS
1920

CAMBRIDGE UNIVERSITY PRESS
Cambridge, New York, Melbourne, Madrid, Cape Town,
Singapore, São Paulo, Delhi, Mexico City

Cambridge University Press
The Edinburgh Building, Cambridge CB2 8RU, UK

Published in the United States of America by Cambridge University Press, New York

www.cambridge.org
Information on this title: www.cambridge.org/9781107646902

First published 1920
First paperback edition 2013

A catalogue record for this publication is available from the British Library

ISBN 978-1-107-64690-2 Paperback

PREFACE

I DESIRE to acknowledge my indebtedness to the admirable articles in the *British Association Guide* on the Geology and Natural History of the district. With regard to the illustrations, most of them are here published for the first time. Many are reproduced from my father's negatives; these are specified in the list, where the sources of all the illustrations are acknowledged. I wish to thank all those mentioned, as well as my wife, for their valuable assistance.

G. N. PINGRIFF

July 1920

CONTENTS

ILLUSTRATIONS

MAPS AND PLANS

The illustrations on pp. 3, 61, 82, 96, 103, 118, 157 are reproduced from photographs by the author; those on pp. 5, 20, 28, 34, 54, 55, 57, 73, 87, 121 from maps and diagrams specially drawn by the author; those on pp. 7, 12, 22, 42, 59, 89, 102, 105, 106, 108, 110, 120, 140, 149, 154, 159 from photographs by Mr W. Pingriff; that on p. 9 from a photograph by Mr W. J. Harrison; those on pp. 15, 79 from photographs by the Geological Survey; those on pp. 18, 35 from photographs by Dr F. J. Allen; that on p. 25 from a photograph by F. Frith & Co. Ltd., that on p. 31 from a photograph by Mr E. E. Lowe; those on pp. 36 and 77 from photographs lent by Prof. W. W. Watts; those on pp. 92, 94, 95, 99, and 125 from photographs by the Rev. Wallace Watts; those on pp. 44, 63 from photographs by the Sport and General Press Agency, Ltd.; that on p. 49 from a diagram prepared by Dr H. R. Mill; those on pp. 67, 69 from photographs supplied by N. Corah & Sons; those on pp. 71, 76 from originals supplied by the Croft Granite, Brick and Concrete Co. Ltd.; those on pp. 85, 97, 117 from photographs by Mr H. Pickering; those on pp. 109, 112, 113, 137, 155 from photographs by Valentine & Sons, Ltd.; that on p. 115 from a photograph by Miss Spencer; those on p. 123 from photographs by Mr F. E. Pingriff; that on p. 128 from an original supplied by Wakefields, Ltd.; those on pp. 129, 130 from photographs supplied by the Midland Railway Co.; that on p. 134 from a photograph by Mr G. Clarke-Nuttall; and those on pp. 139, 142, 144, and 145 from photographs by Emery Walker, Ltd.

1. County and Shire

ENGLAND was first divided into shires during that early period of our history in which the Saxons and the Danes were fighting for supremacy in the land. The word *shire* is of Anglo-Saxon origin, being derived from a verb—*sceran*—meaning "to cut," and it is thus a portion of land *shorn*, or cut off, from some larger area. This dividing up of the land was probably undertaken for purposes of military organisation. It appears to have been carried out, at any rate so far as our own county is concerned, in the early part of the tenth century, during the reign of Edward the Elder, son of Alfred the Great. Although a few of our present English counties, such as Essex or Kent, are the survivals of ancient kingdoms and thus came much earlier into existence, it is not until the end of the tenth century that we find any record of the Midland shires.

At the Norman Conquest, about a century and a half later, the old divisions were, on the whole, fairly closely adhered to. The Normans, however, substituted their own word "county" (*comté*) for the English word "shire"; but in their original meanings the two words do not exactly correspond, for while a shire is simply a share or division of the land, a county is a portion of the land under the rule of a *Comes* or *Comte*, who corresponded more or less to the Saxon *Eorl* or *Thane*. Since Norman times changes in the county boundaries have taken place, but they have been for the most part comparatively insignificant, so that we may fairly say that our modern English counties are substantially the original pre-Norman divisions.

When, however, we come to consider why the English counties have their present boundaries, and what guided the first dividers of the land in shaping them, we find ourselves on more difficult ground. In the days of the Saxon heptarchy—say towards the end of the ninth century—the land was divided into kingdoms (see map on page 54) of which the boundaries were never fixed for any appreciable time, and the division into shires was of course unknown. There arose, however, probably about this time, a division known as the *Hundred*,[1] and it was by the grouping together of these Hundreds that the Midland shires were built up, the shire name in most cases, as in that of Leicestershire, being taken from that of the principal town in the Hundreds concerned. If the outermost Hundreds bordered on a river which could not easily be forded, this river would remain a county boundary, but if they had irregular boundaries or outlying districts then these irregularities and outlying portions would also appear in the county boundaries. So, too, in the case of lands held by religious bodies which it would be inconvenient to have in different counties. In this way we

[1] See p. 132.

can explain many of the irregularities of outline shown by the English counties.

The town of Leicester, from which the county took its name, is of much greater antiquity than the county itself—indeed the name clearly indicates that the place

Ancient Earthworks at Ratby, near Leicester

(Looking through a gap in the "vallum")

existed in Roman times, although it was known to the Romans by another name. At the time of the Domesday survey (1086) the county was referred to as Ledecestrescire, and in other old records the name appears in the following forms, Lege-cestria, Legeocester, and Leger-ceaster. The termination of all of these is evidently derived from the Latin *castrum*, a camp, which has survived in these and various other forms, such as *-chester*, *-cester*, *-caster*, and *-castor*. The first part of the name is in all probability a corruption of the word Leir or Leire, the old name for the river, on the eastern side of which the "castrum" was situated, and now known as the Soar. The ancient name, which is said to have been taken from that of the Lœgre (afterwards Loire) in Gaul, has survived in the name of the small village Leire in the upper valley of the river, and it is interesting to note that in the case of the village also old records give the name of the manor as Legre and Leyre.

The Roman name for the town—Ratae—is in all probability a latinised form of the still older British name Rhage, which may possibly be related to the British word *rath*, signifying a cleared space used as an encampment. The plural form of the Latin name seems to indicate that more than one of these ancient encampments existed in the neighbourhood. Ratby, a few miles away, was probably another, and at this village ancient earthworks may still be seen.

2. General Characteristics. Position and Natural Conditions

Leicestershire, in spite of the fact that it has no very marked natural boundaries such as mountain ranges or arms of the sea, is in many respects one of the most compact of all the English counties. If we could imagine it completely isolated from the important neighbouring counties on all sides it would make an admirable county-state, either alone, or together with

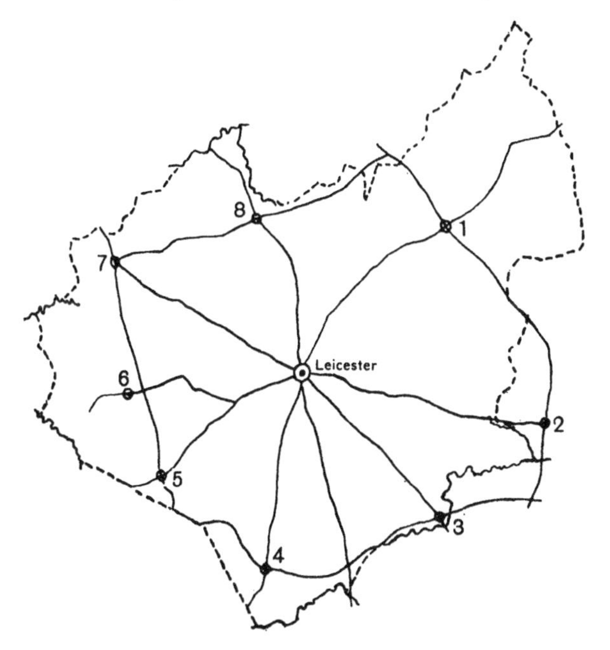

The Market Towns around Leicester

1. Melton Mowbray. 2. Uppingham. 3. Market Harborough. 4. Lutterworth. 5. Hinckley. 6. Market Bosworth. 7. Ashby de la Zouch. 8. Loughborough.

the tiny county of Rutland which adjoins it on the east.

This compactness is well shown in the annexed map of the market towns. The county town is situated almost exactly in the centre of the county; it is in all respects a true capital, the next largest town, Loughborough, having only about one-tenth of its population, and it is surrounded by a ring of lesser market towns. The main roads radiate from the centre like the spokes of a wheel from the hub, while the rim of the wheel is completed by a series of secondary roads which only to a very slight extent cut into the bordering counties. With one important exception, which will be mentioned later (page 122), the map shows practically all of the important roads of the district.

Leicestershire includes an almost complete river basin—that of the Soar—which flows right across the county from south to north, dividing it into two approximately equal areas. These are in some ways very unlike in their natural characteristics. Their geological formations are quite distinct, and the same is consequently true to a great extent of their soils and scenery, while to a lesser degree the difference may also be seen in the occupations of the inhabitants and in the characteristic architectures of the two portions. The eastern side of the county is almost entirely agricultural; the soil here is highly fertile and forms excellent grazing country, and there are no waste lands such as heaths, moors, or bogs; even the highest points of the picturesque hills and rolling country being, as a rule, enclosed and used for grazing purposes. The western half of the county, on the other hand, although still to a great extent agricultural, has other, and in many places more predominant interests. Here, chiefly in the north-western corner, are to be found stone-quarries and coal-fields, which are especially important owing to the fact that they are among the nearest

Typical rolling country of East Leicestershire

of their kind to London and the south-eastern counties; and here also occurs a remarkable region known as the Charnwood Forest, described more fully in a later section, where the scenery is altogether unlike that of any other portion of the county.

The country, then, on the east and south-east is entirely agricultural, that on the west and north-west agricultural and mining, but the towns—Leicester itself, as well as most of the lesser towns—are best described as industrial, though in some cases the manufactures are only of comparatively recent growth. The tendency here, as in many other parts of England, seems to be distinctly towards an increase in the industrial population at the expense of the agricultural: some of the smaller towns, such as Loughborough and Hinckley, have of late years become quite important manufacturing centres, a result which was perhaps only to be expected in view of their proximity to the Leicestershire coal-fields, as well as to those of the neighbouring counties of Derbyshire, Warwickshire, Staffordshire, and Nottinghamshire.

In position Leicestershire may be regarded as almost the centre of England. It is of interest, therefore, to try to discover in what way its central position has affected its general characteristics.

Being a long way from the sea, having no great natural waterways, and not lying very near either to the metropolis or to the great manufacturing districts further north, it seems to have come to depend largely upon its own resources, and this probably has added to that compactness already referred to. Yet it is in reality not at all cut off from the rest of the country, as is clearly shown by the fact that ever since Roman times Leicester has occupied an important position on the main routes up and down or across the country. At the

Broombriggs Hill: a typical view of the Charnwood Forest country

present time two of the great railways running northward from London pass through the county town, while feeders come in from each of the two other great lines lying respectively east and west of the county. Lastly, its position in the central uplands of England has doubtless contributed both towards its equable climate and to the multiplicity of small streams by which it is watered, with its consequent agricultural prosperity.

3. Size. Shape. Boundaries

If an average be taken of the size of all the English counties, Leicestershire is found to be slightly short of it, the administrative area to-day being given as 530,642 acres, or almost exactly 800 square miles. Its area, compared with that of the whole of England, is shown by the figure at the end of the volume.

The shape of the county is roughly that of a pentagon, with a broad bay entering the northern side, while the north-eastern corner extends outwards as a large promontory into the flat lands of the neighbouring counties of Nottinghamshire and Lincolnshire. The greatest length of the county (45 miles) is that from this north-east corner—Three Shires Bush—to the southern apex near Rugby, and its greatest breadth, that from the most easterly point, where it meets the counties of Lincolnshire and Rutland, to the most westerly, where lies its junction with Staffordshire, Derbyshire, and Warwickshire, is nearly as much—41 miles.

Far more interesting, however, than the above figures is a consideration of the boundaries themselves as they would appear to a person setting out upon an actual circuit of the county. Let us suppose that our traveller begins his journey at the north-eastern extremity. He has passed the night at the little town of Bottesford in the Vale of Belvoir, and a three-mile walk to the north along the Newark road has brought him to Three Shires Bush, his actual starting point. His journey will at once resolve itself into a cross-country scramble, and in order to follow the exact boundary for even a short distance, a 6-inch Ordnance Survey map will be necessary. At times his path, which leads him to the south-west, will follow the hedgerow or country lanes, at others he must travel by the sluggish streams and dykes found in this district, and now and again he will pass along a stretch of the main highway. Journeying thus our traveller might reasonably expect to accomplish, in his first day, some 10 or 12 miles, which would bring him from the flat country in which he started to the neighbourhood of Upper Broughton or Dalby on the Wolds, the latter one of the most beautifully situated villages in the county. Thence a good road across the hilly country for the next 10 miles or so—which, though not actually constituting the county boundary, follows it sufficiently closely—leads him to the now important town of Loughborough, lying about 2 miles within the county. From here or close by he could travel the next 15 miles or so by water, for the boundary now follows the course of the river Soar down to its confluence with the

Trent, and thence passes up the latter river to the beautiful eminence on which stands the now well-known Donnington Hall, about 2 miles west of the ancient little town of Castle Donnington.

The boundary here leaves the Trent and leads to the

A Leicestershire Lane, showing the well-wooded hedgerows

south-west, a most difficult and ill-defined yet interesting cross-country stretch of about 12 miles, to Chilcote, the most westerly village in the county, and thence again towards the south-east for another 6 miles, to the picturesque little river Anker, flowing past Atherstone, just across our border. Here the fine Roman road—Watling Street—forms the boundary, and leads our traveller with undeviating course, first to its junction with the Fosse Way at High Cross—the Roman station of Venonae—and thence to the southernmost corner of the county, in the parish of Catthorpe. This is in the river basin of the Severn, and the boundary now turns north-east, and follows the upper waters of its tributary, the Avon, for 7 miles or so. Next, the narrow watershed between this river and the Welland is crossed, and, passing Market Harborough, our traveller traces the boundary down the course of the latter river, with Northamptonshire on the right bank, for no less than 17 miles to the Rutland border, which is still a water boundary, formed by a Welland tributary, the Eye Brook. Following this up for about 6 miles to the north-north-west, as far as the main road from Leicester to Uppingham, he now turns north to cross the high rolling uplands of East Leicestershire. At Whissendine, Rutland makes a considerable bay into the county, and the boundary bends eastward to beyond Wymondham, when it once more turns sharply to the north, and follows the road for 10 miles or more to near Harston, whence, passing close to Belvoir Castle, and still holding its northerly course, it brings the traveller back to his starting point near Normanton.

It is interesting to note that, with a total perimeter of about 164 miles, no less than 70 miles of this is composed of rivers and brooks—the Trent, Soar, Welland, Avon, Eye Brook, Anker, Mease, and others, each taking a part in fixing the boundary. For 42 miles the county demarcation is by roads or lanes, while for the remaining 52 miles or so the dividing line is arbitrary and across country. The only important straight boundary is that formed by the Roman road, Watling Street, a fact which clearly illustrates that even before our counties or the Hundreds of which they are composed came into existence the road was there.

In the year 1888, when so many of the county boundaries were rectified, Leicestershire underwent certain alterations, especially on the Derbyshire border, where various isolated portions of that county were incorporated.

4. Surface and General Features

As already stated, Leicestershire is divided into two rather dissimilar portions by the river Soar. We shall now describe the surface features of these different areas in rather more detail, and in a later chapter endeavour to discover how far these features can be accounted for by geological conditions.

In passing through a district by train we may or may not be able to obtain a correct idea of its physical geography, but a surer method is to view the country from its most elevated points, and to note, by the aid of a compass, their relative positions. We may begin by the high ground of the north-eastern portion of the county —the district to the north of Melton Mowbray. There are many excellent view-points in this district, such as Belvoir Castle, the high ground in the neighbourhood of Harston or Croxton Kerrial, or Green Hill near to

View near Wartnaby

Wartnaby. The illustration here given shows a fairly typical stretch of country, though many finer and more extensive views may be obtained across the Vale of Belvoir to the east, but as usual in such cases photography fails to do justice to the actual effect. It is not that the land in this part is really of very great height, for it seldom reaches more than about 550 feet, but the surrounding country, which to a great extent lies over the borders of Nottinghamshire and Lincolnshire, being exceptionally low, the same effect is produced, and we can readily see, as the geologist would tell us, that we are upon an escarpment of the hard rocks, which has been formed in a past age when the sea, covering what is now the plain below, wore back the hills to their present position.

The streams on the northern wolds and uplands which we have been considering run mostly to the north, but if we now journey to the south we shall soon discover that the streams accompany us, and that the land slopes down to the valley of the Wreak. Crossing the rich lands of this district, which the Danish settlers of a thousand years ago found so much to their liking, we find the land rising again in the uplands of east Leicestershire, and have no difficulty in again obtaining suitable view-points. Burrow Hill, where was an ancient British camp, Robin-a-Tiptoe, Whatborough Hill, or many other points in the neighbourhood of Tilton, are good examples. We are now in the famous hunting country of the Quorn, the Fernie, and the Cottesmore packs, a country of steep hills and dales and broad views, of fine old churches and houses built of the local marlstone, a country of grazing land, where a ploughed field is a rare sight, but, alas, a country of ever diminishing population.

But now, instead of continuing our journey further south, let us start anew at the north-west. Here are the highest points of the county, indeed of this part of England, and consequently still more extensive views may be expected. From Bardon Hill (912 feet) one may see the Malvern Hills to the south-west, the Derbyshire heights to the north, and, it is said, even Lincoln Cathedral to the east; and from Beacon Hill (818 feet), a few miles away, the view is perhaps finer, though somewhat less extensive. But from either of these points we shall soon realise that we are on entirely different ground from that on the eastern side of the county. This district, known as the Charnwood Forest, though long since deforested, comprises a comparatively small area (about 50 square miles) of a remarkable nature. A good idea of the nature of the country is given by the illustration on p. 9. Islands, as it were, of some of the most ancient rocks known, are thrust up like a miniature mountain range through a surrounding sea of later-formed rocks. Everywhere in this district the granite or other rocks of almost equal antiquity may be seen washed bare of the covering mantle of fertile soils which still fill the valleys, thus giving rise to a vivid contrast between steep bracken-covered crags and cultivated fields. The most rugged part of the whole district is perhaps that in the neighbourhood of High Tor and Pedlar Tor, near Whitwick. Bradgate Park, the home of Lady Jane

Grey, forms the southern corner of the Forest, and is only about 5 miles from Leicester.

To the west of the Charnwood Forest, in the neighbourhood of Ashby de la Zouch, lies the Leicestershire coalfield which, as one would expect, is somewhat thickly

Volcanic Rocks of the Charnwood Forest Region

populated, but does not possess natural features of any particular interest. Thence, as we pass to the south, the country becomes more and more agricultural, and the gently undulating surface of the upper basin of the Soar merges almost imperceptibly into that of the Avon, or of the Welland to the south-east. Here the country is somewhat lacking in good view-points, the highest ground being in the neighbourhood of Knaptoft and Mowsley, where a main road from Leicester to the south reaches a height of 575 feet. This district is connected by a ridge of moderately high ground with the uplands of East Leicestershire, which have already been discussed.

The soil of the southern, as also of much of the eastern portion of the county, is for the most part a heavy clay, the ground being thickly overspread by the—geologically speaking—comparatively recent deposits of the glacial epoch. There is practically no barren land in this part of the county, which is mostly in pasture, though ploughed fields are also common. We have thus now passed from the agricultural country of the east and north, glanced at the granite-producing and mining districts of the north-west and west, and returned to the agricultural south and south-east, and in so doing have reviewed most of the important physical features of the county, with the exception of the rivers and streams. It remains, therefore, to study these and one or two interesting points in connection with them before proceeding to a consideration of the geology of the district.

5. Watersheds and Rivers

Leicestershire cannot boast of any very imposing rivers, but it is closely and uniformly covered by a network of small streams, mostly flowing into one main artery—the Soar—which itself runs into the Trent on the northern boundary of the county. These numerous streams have contributed greatly to the agricultural prosperity of the county.

On looking at the map, in which watersheds are shown

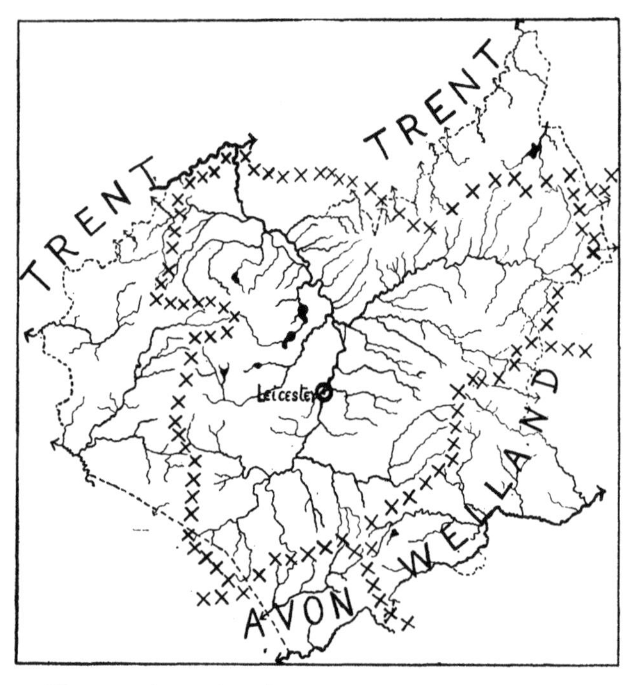

Rivers and Brooks of Leicestershire, showing extent of the Soar Basin

by a series of crosses, it will at once be noticed that by far the greater part of the county lies within the basin of the river Soar, which drains into the sea by way of the Humber. From Leicester to the Humber the distance by water is approximately 110 miles, and the fall of level about 180 feet, hence the current cannot be expected to be of great strength. The valleys of the Soar and its principal tributary the Wreak, which flows from the north-east, are well marked, as is shown by the orographical map at the beginning of the book ; they are somewhat liable to floods, though not to such an extent as are the valleys of the Welland and the Nene, further to the south ; they are extremely fertile, and exhibit a good deal of quiet beauty. The Soar rises in several small streams on the south-eastern border of the county. Its principal tributaries are the Wreak, with its feeders the Eye and the Queniborough Brook, the Billesdon Brook (also known as the Sence), and the Anstey Brook. The Wreak flows into the Soar from the north-east ; it rises to the north of Melton Mowbray, and is a stream of considerable magnitude on reaching the Soar near Cossington, about 5 miles to the north of Leicester. The Billesdon Brook, bringing water from the high ground near to Billesdon, after bending sharply to the west near to Great Glen, flows for the last 6 miles of its course parallel to the Grand Junction Canal (see Chapter 19), though at a lower level. The Anstey Brook is one of a series of streams which gather their water on the high ground of the Charnwood Forest.

Outside the Soar basin there are portions of other river basins which are of considerable interest. In the north-east is a small area, including the picturesque reservoir at Knipton, which sends water direct to join the Trent near to Newark, and on the western side of the county is a larger area draining into the upper Trent. This latter district includes the bright and clear little river Sence, abounding in water-rats and voles, which joins the Anker near to Atherstone. The Anker, for a short distance a county boundary, eventually delivers its

The Soar, near Leicester

waters into the Trent by way of the Tame, which it joins at Tamworth.

So far we have been entirely in the Trent basin, but to the south of the county is a small area draining into the Severn by way of the Avon, which, as already mentioned, forms the southern boundary of the county for a distance of some 6 miles. This district also includes the little river Swift, flowing through Lutterworth, which place was for many years the home of John Wycliffe, whose bones were thrown into the Swift after being disinterred in 1428. In the words of Thomas Fuller, written in the seventeenth century—" This brook has conveyed his ashes into Avon, Avon into Severn, Severn into the Narrow Seas, they into the main ocean, and thus the ashes of Wycliffe are the emblem of his doctrine, which is now dispersed all the world over."

In the south-east of the county is a district draining into the Welland, which itself rises within the county near Husband's Bosworth, and together with its tributary the Eye Brook forms an important county boundary. Lastly, a very small area near to the Rutland border drains into the Witham.

It will thus be seen that Leicestershire occupies an important position with respect to the main watersheds of England. At Knaptoft, in the south of the county, in a region of not more than about a square mile in area, there are streams which deliver their water to points as far distant one from another as the Humber, the Wash, and the Severn Estuary; this will easily be found in the river map of this chapter.

6. Reservoirs and Water Supply

The principal sheets of water in Leicestershire are in that part of the Soar basin which comprises the Charnwood Forest district. Of these only one is of natural origin—that known as Groby Pool—the largest natural lake, if such it may be called, in the county. Considerably larger, however, are the artificially constructed reservoirs in the neighbourhood, and these, which are not without interest, may be named in the order of their construction. They are all, from a scenic point of view, well worth visiting, and for many years have afforded the inhabitants of both Leicester and Loughborough an adequate supply of pure water. The oldest, situated at Thornton, was constructed in 1854, and gathers its water from the western slopes of the Charnwood area. It has a capacity of some 330 million gallons, and being at a height of about 200 feet above Leicester, the water easily gravitates to the town. The next, Cropstone, finished in 1870, has a greater capacity—560 million gallons—and gathers its water from a large area in the central part of the Forest ; it is very picturesque, and with the rugged hills in the background, makes as fine a landscape as may be seen in central England. The third, at Swithland, completed in 1896, has a capacity of 490 million gallons, but, it is commonly understood, has not altogether fulfilled the expectations of the promoters of the scheme. The water from Swithland as well as from Cropstone has to be pumped to Leicester. The fourth reservoir, in the north of the district, was completed in 1906, and is known as the Blackbrook Reservoir ; it supplies the neighbouring township of Loughborough. The small reservoir at Saddington in the Welland basin merely supplies water to the neighbouring canals, while Knipton Reservoir, in the valley of the Devon, serves the Grantham Canal in the same way.

The three reservoirs above, however, were found to be insufficient to meet the ever-increasing demands of the

Knipton Reservoir and Belvoir Castle

population of Leicester, and in 1899 a new Act of Parliament, known as the Derwent Water Scheme, was passed, whereby Leicester was to obtain an additional supply from the valley of the Derwent, in the north of Derbyshire: it shares this privilege with the important neighbouring towns of Derby, Nottingham, and Sheffield.

7. Geology

In order to appreciate the full value of a geological map such as that shown at the end of this book, it is first necessary to acquire some slight knowledge of the principles of that branch of science which is known as stratigraphical geology. As probably all our readers are aware, the rocks (a term in geology which includes even soft deposits such as clay) of which the earth's crust is composed are not all of the same age. The oldest of all are those "igneous" rocks, which were formed when a crust first began to appear around the previously fluid matter from which the earth solidified. As time and cooling went on, more and more solid rock made its appearance, until eventually the temperature was sufficiently reduced for water to exist as such on the surface, and the seas came into being. But from the beginning the rains and winds began to carve upon the solid crust from without, while volcanic action and earth movements, which in these earlier days must have been rife to a much greater extent than now, added their disturbing effects from within, and the first formed rocks consequently did not long remain in their original state. The rains and frosts acting upon them dissolved and broke them up little by little, while the waters carried the materials of which they were composed to the seas and deposited them there. The seas gave place to dry land, and the "sedimentary" rocks made their appearance only themselves to be again carved upon, washed away, or even, over and over again, submerged beneath the oceans of by-gone days. The result of these changes is that now almost the whole of England, to quite a great depth, is composed of rocks which were formed under water. These aqueous rocks, of course, were not all laid down during any one geological age, and rocks deposited in different periods vary much in their general formation, so that by examining the common surface rocks found in a given locality a geologist can say to which particular period these rocks belong. He would, for instance, be able to state that the rocks on the western side of our county belong to a rather earlier period than those on the south-east, while the latter he would know are somewhat older than the limestone rocks found in many parts of Northamptonshire, and these again older than the common rocks found still further to the south-east. He would know this because, as a rule, the rocks in the first-mentioned situations are found, by boring deeply enough, also to occur beneath those mentioned as belonging to a more recent period. He would explain it in a way that should be made clear by the map and accompanying section taken across the county on a line joining London and Leicester. The successive strata, which,

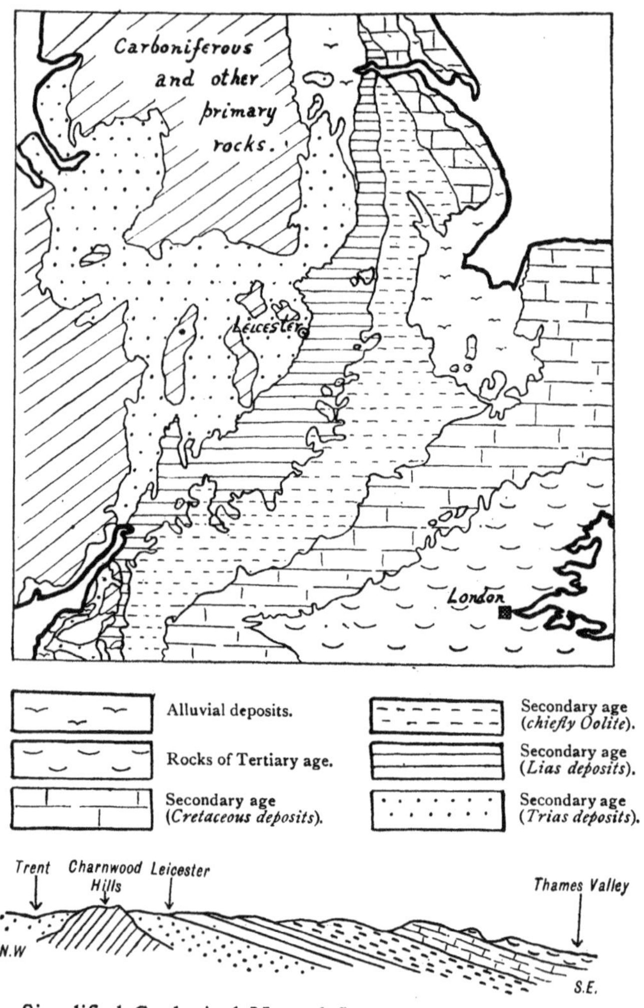

Simplified Geological Map of Central England, together with section from S.E. to N.W.

taking first the most recently formed, run from south-east to north-west, were originally all horizontal, but have in some way become tilted to the position shown; the uplands, formed as a result of this, have suffered most from erosion, and older and older strata consequently appear on the surface as we journey to the north-west. If the section be compared with the geological map of the country in which it occurs, the relationship between the two will be at once seen.

The various systems of rocks which would be passed through in a boring at a place where all the known strata had been deposited, are shown in the more complete table on p. 30. Such a boring might be 20 miles or more in thickness, and of course has never been realised: it is only mentioned here in order to indicate the relative positions in it of those systems of rocks which may be found on the surface (*i.e.* "outcrop") in Leicestershire. These are underlined in the table, and will be referred to again. All of the systems except the Miocene are found in the British Isles.

Each of the epochs named, except the very earliest, had its own characteristic forms of life, and owing to the fact that remains of many of the ancient species have been preserved as fossils in the contemporaneous strata, we are able to form a pretty close idea of the actual biological conditions in some of these past ages. Thus the fossils of the Carboniferous period include numerous remains of tree-ferns (*Lepidodendron*, an order now represented by a small moss, but which then included trees 50 feet

	Names of Systems	Subdivisions	Characters of Rocks
TERTIARY	**Recent** **Pleistocene**	Metal Age Deposits Neolithic ,, Palaeolithic ,, Glacial ,,	Superficial Deposits
	Pliocene	Cromer Series Weybourne Crag Chillesford and Norwich Crags Red and Walton Crags Coralline Crag	Sands chiefly
	Miocene	Absent from Britain	
	Eocene	Fluviomarine Beds of Hampshire Bagshot Beds London Clay Oldhaven Beds, Woolwich and Reading [Groups Thanet Sands	Clays and Sands chiefly
SECONDARY	**Cretaceous**	Chalk Upper Greensand and Gault Lower Greensand Weald Clay Hastings Sands	Chalk at top Sandstones, Mud and Clays below
	Jurassic	Purbeck Beds Portland Beds Kimmeridge Clay Corallian Beds Oxford Clay and Kellaways Rock Cornbrash Forest Marble Great Oolite with Stonesfield Slate Inferior Oolite Lias—Upper, Middle, and Lower	Shales, Sandstones and Oolitic Limestones
	Triassic	Rhaetic Beds Keuper Marls Keuper Sandstone Upper Bunter Sandstone Bunter Pebble Beds Lower Bunter Sandstone	Red Sandstones and Marls, Gypsum and Salt
PRIMARY	**Permian**	Magnesian Limestone and Sandstone Marl Slate Lower Permian Sandstone	Red Sandstones and Magnesian Limestone
	Carboniferous	Coal Measures Millstone Grit Mountain Limestone Basal Carboniferous Rocks	Sandstones, Shales and Coals at top Sandstones in middle Limestone and Shales below
	Devonian	Upper Mid Lower } Devonian and Old Red Sandstone	Red Sandstones, Shales, Slates and Limestones
	Silurian	Ludlow Beds Wenlock Beds Llandovery Beds	Sandstones, Shales and Thin Limestones
	Ordovician	Caradoc Beds Llandeilo Beds Arenig Beds	Shales, Slates, Sandstones and Thin Limestones
	Cambrian	Tremadoc Slates Lingula Flags Menevian Beds Harlech Grits and Llanberis Slates	Slates and Sandstones
	Pre-Cambrian	No general classification yet made	Sandstones, Slates and Volcanic Rocks

high), and the coniferous trees, as well as many species of corals, echinoderms, molluscs, and fishes. And again, the later rocks of the Lias include among their many characteristic fossils such forms as *Gryphæa* (very common in Leicestershire), *Lima* and Ammonites, as

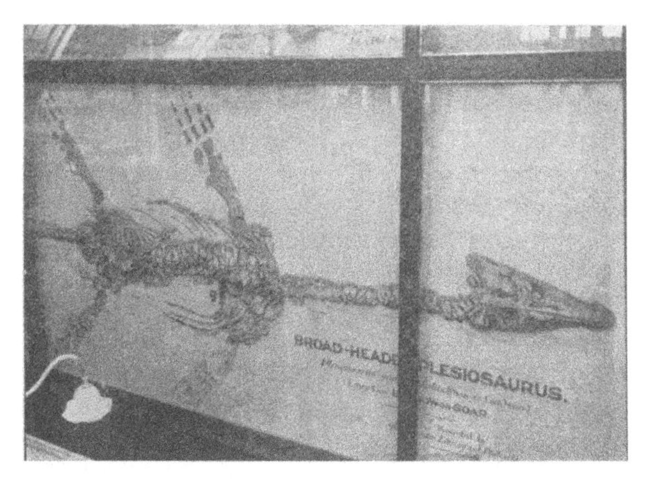

Plesiosaurus macrocephalus from Barrow-on-Soar

well as remains of the remarkable species of reptiles known, from their relationship to the lizards, as Saurians; many of these have been found in the clayey limestone beds of the Lower Lias at Barrow-on-Soar. It is interesting to notice, however, that the corals, so plentiful in earlier beds, are almost absent from those of the Lias—the waters of this period were evidently too muddy for them to thrive.

From the map on page 28, and the key beneath, we can obtain a good general impression of the geology of Leicestershire, and it will be found instructive to compare this with the more complete map at the end of the book. We see that the rocks on the eastern side of the county belong for the most part to the Lias, while those on the western side come mainly in the Trias, but include certain small areas composed of much older rocks: we also see that nowhere must we expect to find either chalks or the comparatively recent deposits of the Tertiary period. It will be sufficient for our present purpose to consider only the sub-divisions of such of the above systems as outcrop in Leicestershire. The chief sub-divisions of the Lias and Trias, with the immediately over-lying strata, are shown in the table on p. 30. The Oolite is of little importance, and is represented only by a few patches of its oldest rocks, capping the hills in the north-east, round Croxton Kerrial and Saltby, and a smaller patch further south, near to Medbourne. The Upper Lias is of not much more importance, but forms a rather steep bank, running close to the eastern border of the county, and not stretching far into it before giving place to the outcrop of the Middle Lias, which is of much more interest. This is easily recognised wherever the marlstone rock may be seen free from its covering of Drift (to be mentioned later): it forms a series of fine escarpments overlooking the Lower Lias plains further to the west. These escarpments may be well seen at several points near to Tilton, at Burrow Hill, and on the line running from Ab-Kettleby, north of

Melton Mowbray, to Belvoir Castle, which lies still further to the north. The rock is of a deep rust-colour, and indeed contains much iron. It is very porous, but notwithstanding this, is a useful building stone. At its base numerous springs arise. The Lower Lias, which comes next as we travel west or north-west, covers a large area, reaching almost to the valley of the Soar. It consists, for the most part, of clayey limestones, which are worked at Barrow-on-Soar and a few other places, as well as of shales and pure clays : it rises to a fair altitude in the neighbourhood of Six Hills, to the north-east of Loughborough, and forms the wold country, which was referred to in Chapter 4.

The Rhaetic beds, Keuper marls and sandstones, and Bunter beds of the Trias—the Keuper marls being of most importance—come next, and cover three-quarters of the western half of the county, their relative positions being shown in the coloured map. These triassic sandstones and marls form different soils from those of the Lias. They appear to be on the whole less suitable for cereal crops, but perhaps more so for fruit trees, but to a great extent the triassic rocks, like those of the Lias, are covered by Boulder clays and other Drift deposits (*cf.* also p. 37).

In the north-west of the county the Keuper marls and sandstones give place to still older rocks belonging to the Carboniferous system. These lie in the neighbourhood of Ashby-de-la-Zouch, and, owing to the fact that their upper beds contain numerous seams of coal, they are of great economic importance. They are again referred to in the chapter on Mines and Quarries, and therefore need not be further discussed here.

But in addition to the Carboniferous, tiny patches of Permian, and the very complete series of later secondary rocks found in the county, Leicestershire offers unique opportunities for the study of an interesting group of much older rocks. These underlie the Charnwood Forest district, and have probably attracted more workers than

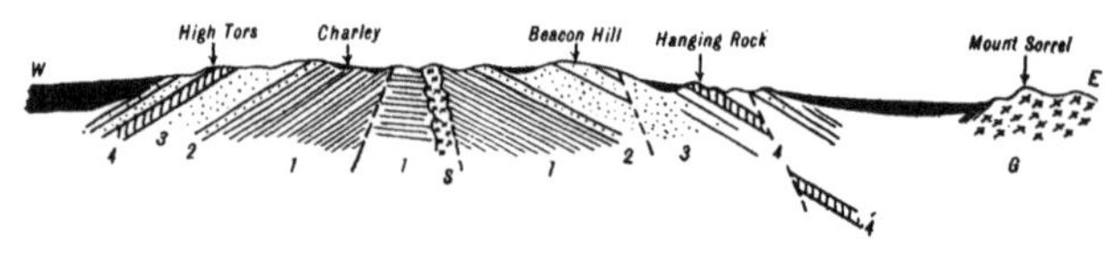

1. Hornstone and grits. 2, 3. Felsitic agglomerate, etc. 4. Slate agglomerate, followed on E. side by true slate. 5. Syenite. 6. Granite. Black = Triassic marls.

Simplified section across Charnwood District

any other district of equal area in the country. They are of great age, and it has been suggested that a new Pre-Cambrian system—the Charnian system—should be instituted. The rocks themselves are mainly of volcanic origin, but the material of which they are composed appears to have been often laid down under water. The actual "lie" of the rocks, as well as their relationship to the overlying triassic rocks, is shown in the generalised section through the district, represented by the sketch here given. It is clearly seen that the old rocks rise up through the triassic strata like an island rising from the depths of the sea, but the section does not show how the valleys have been filled by the newer rocks, and how these have been cut out again by the present streams. The order of the rocks has been very thoroughly worked out, but it must suffice here to name a few of the chief varieties, such as are shown in the sections or illustrations of the present volume. The Felsitic agglomer-

Volcanic Agglomerates of the Charnwood District

ates and the Hornstones are probably the oldest of the volcanic rocks; then after various other conglomerates and grits come the Swithland slates, which have been worked at various quarries in the district. A very fine-grained slate, found at Whittle-hill, has long been famous for the excellent hones which it furnishes. The original igneous rocks, also found in the district, include the porphyroid rock of Bardon Hill as well as syenites and diorites, which are extensively quarried for road metal, both in the Forest itself and at various localities in the south of the county. Lastly, there is a famous pink or grey granite, which comes to the surface at Mountsorrel, somewhat to the east of the Forest proper.

Weathered Mountsorrel Granite

The geology of the district is rendered difficult by the numerous "faults" which have occurred, dislocating the beds from their normal position, as well as by the fact that in studying volcanic and igneous rocks such as these, no light can be obtained from the fossil remains, for there are none, unless they are to be seen in the few "worm burrows" which have been recorded. Summing up the geology of the whole district, Professor Watts describes it as a " Triassic landscape, preserved fossil in the marl, and developed by recent denudation for our study," but of course even now it is only partially laid bare.

The Drift deposits, which were mentioned earlier in the chapter as covering large areas of the county, remain to be considered. It is believed that the various clays, gravels, and sands of which the Drift is composed were in quite recent geological times spread over the older rocks by immense glaciers. Scratches on the rocks and other records clearly show that at least one of these glaciers reached as far south as the Thames Valley. On their slow journeys to the south they wore off materials from the rocks further north and carried these with them, until the rising temperature melted the ice, and allowed the rocks or their debris to be spread over the ground beneath. The glacial deposits of Leicestershire appear to be of two distinct ages—the older deposit being from the north-west, the other, which includes occasional masses of oolite limestones, being more from the east or north-east. The chief glacial deposits in Leicestershire are Boulder clays: this clay can be well seen in the illustration on p. 79, where it forms a heavy covering over the middle Lias ironstone. Glacial deposits have been found to a thickness of more than 700 feet, and are thickest in the centre and south of the county, being almost absent in the north between the Charnwood Forest and the Trent. At the close of the Glacial Epoch many new species of animals, including some of the larger mammals, such as the rhinoceros, wild ox, the mammoth and other elephants, evidently roamed over the country, as their remains testify, and at about this period also man made his first appearance.

8. Natural History.

During the different epochs through which our earth has passed, that portion of England which we now call Leicestershire has at certain times been dry land, and at others fathoms deep beneath the seas. It is well known, indeed, that the British Isles did not exist as such until some time after man had spread over this part of western Europe. Even now the real Atlantic Ocean may be said to begin some fifty miles or so to the west of Ireland, where the depth increases rapidly from some hundred fathoms only to more than ten times as much, while on the other hand, the North Sea is scarcely anywhere deep enough to submerge St Paul's Cathedral.

It would be expected therefore that, the separation of Britain from the mainland being of such recent geological date, our country would be peopled by practically the same species of plants and animals as are to be found on the continent of Europe. As a matter of fact, at the present day a large number of European species are entirely absent from Great Britain, and still more from Ireland, a result due primarily to the conditions existing during the Glacial Epoch of Pleistocene times. After the ice sheets from the north had swept away the once abundant fauna and flora of the country, re-population took place on the return of more genial climatic conditions. This began from the direction of our connection with the Continent on the south and east, but before the species had all re-established themselves the sea had cut us off.

Of the present animal and plant population of Great Britain, Leicestershire can boast but a moderate number of the total recorded species. This of course is not due to any isolation from the rest of the country, but rather to the fact that our county cannot offer suitable dwelling-places in sufficient variety. It has, for instance, no coast-line, no large lakes, fens, or swamps, no moorland, and no mountains, hence the many kinds of plants and animals which in course of time have become adapted to such situations are absent from our local flora and fauna. Further, of those species which within comparatively recent times dwelt with us, many have doubtless disappeared even during the past hundred years or less, owing to the excessive draining of the land and the high degree of cultivation which has been introduced. The flora of the Charnwood Forest district must, within historic times, have been more extensive than it now is, though even to-day it includes a great number of interesting plants which occur nowhere else in the county. It is said that when Lady Jane Grey lived at Bradgate Hall, "a squirrel might be hunted six miles without touching the ground," and a traveller might journey "from Beaumanor to Bardon [also about 6 miles] on a clear summer's day without seeing the sun."

Of plants reported by the older botanists in this district many, such as the sundews (*Drosera*) and the black spleenwort (*Asplenium adiantum nigrum*) have ceased to exist since the drainage of the bogs and cutting-down of the timber, but a few interesting species, which still linger, may be mentioned: *e.g.* the stork's bill (*Erodium cicutarium*), the buckthorn (*Rhamnus Frangula*), the burnet rose (*Rosa spinosissima*), the silvery cinquefoil (*Potentilla argentea*), the purple loosestrife (*Lythrum Salicaria*), the pennywort (*Cotyledon Umbilicus*), the stinking groundsel (*Senecio viscosus*), the giant bluebell (*Campanula latifolia*), cat mint (*Nepeta Cataria*), the bog pimpernel (*Anagallis tenella*), the snake-root (*Polygonum Bistorta*), the spurge laurel (*Daphne Laureola*), the crowberry (*Empetrum nigrum*), the marsh orchis (*Orchis latifolia*), lady's tresses (*Spiranthes autumnalis*), crow garlic (*Allium vineale*), sweet flag (*Acorus Calamus*), and many others, including numerous species, some rare, of *Carex*, the sedges. The moisture-loving plants named above may almost all be found in close proximity to Groby Pool, which lies just to the south of the Forest district, and is the largest natural sheet of water in the county. The other botanical divisions of the county include few uncommon, much less rare plants: the water violet (*Hottonia palustris*) grows in the Trent, where this river forms the northern boundary, and a few characteristic plants, such as the horse-shoe vetch (*Hippocrepis comosa*) and the small-flowered gentian (*Gentiana Amarella*) have been recorded as natives of the small oolite patches which occur near the eastern boundary. The common foxglove (*Digitalis purpurea*) also is only found wild in those soils which occur on the western side of the Soar, and the wild daffodil (*Narcissus Pseudo-narcissus*) has its home at Harby Hills, and nowhere else in the county. Nevertheless, taken as a whole, the county compares well with its neighbours in the number of native species of plants, some 770 of which have been recorded. Nottinghamshire, perhaps the most similar of all the surrounding counties, has 772, and Lincolnshire, a much larger and maritime county, has some 880. It is, however, difficult to fix the exact number of native species, for many of the now common species have been introduced from outside at different periods. The extension of railways and canals, for instance, is responsible for the scattering of seeds along the lines, and plants which are complete strangers to the county, or even to Britain, are frequently found in the neighbourhood of flour-mills and such places. In 1847 an American water-weed (*Elodea Canadensis*) made its first appearance in this country in a canal near Foxton: now it is found in every canal in the county, and is much too plentiful in most.

Many branches of our local natural history have been left a good deal unworked, such for instance as the molluscs, the crustaceans, arachnida, fishes, amphibians, and reptiles. Of insects, the beetles and butterflies, on the other hand, have been pretty thoroughly investigated by a few enthusiasts, among whom must be mentioned H. W. Bates, the naturalist of the Amazon, and his brother. The fishes are those of the neighbouring counties, and exhibit few peculiarities. Salmon have been occasionally found in the river Soar, and trout flourish in some of the larger and clearer streams, such as that which flows through Bradgate Park.

The birds, similarly, are just those which we should expect to find in an inland county such as ours, which offers such restricted physical conditions, and is not on any particular line of migration. There are some sixty

Swans on the Soar

resident species clearly established, and 220 in all have been met with, including both summer and winter visitors. The raven, which was fairly common sixty years ago, has disappeared. Other birds show considerable increase in numbers, notably the starling and the sparrow, and, in a minor degree, the nightingale and hawfinch also seem to be becoming more numerous. The ring ouzel and the dipper or water-ouzel, which used to breed in the Charnwood Forest district, now only occur as stragglers from Derbyshire. The little owl (*Athene noctua*), as is the case in other parts of England, especially the Eastern counties, is becoming unwelcomely numerous. With regard to the mammals, as with the birds, some of the speoies are becoming, or have already become, extinct. The wild cat, pole cat, and pine marten have disappeared, while such animals as the red deer, the badger, otter, and fox have only persisted owing to the protection which has been extended to them by sportsmen and others. The red deer is indigenous at Bradgate Park. The badger is not uncommonly seen, and several cases have been recorded of this animal dwelling in amity with the fox, while the fox itself is probably more plentiful than in any other shire. Leicestershire indeed having been for many years the most noted of the fox-hunting counties, it may not be inappropriate to conclude this section with a few words relating to the development of this particular sport.

Throughout the middle ages, and indeed down to about the middle of the eighteenth century, although many other forms of the chase were popular in this country, the fox as an animal of venery does not seem to have been considered worthy of attention. About the year 1750 hounds appear to have been first employed to hunt the fox, and in 1782 a pack came into the possession of Hugh Meynell of Quorndon Hall, and he it is who must be considered the father of fox-hunting, though an earlier pack seems to have been in the possession of T. Boothby, of Tooley Hall, also in Leicestershire. Meynell conceived a clear idea of what should be the points of a first-class hound, and proved himself a most successful breeder. The modern English foxhound, with his broad, but not too large head, and his

The Quorn Hounds

muscular frame is as near perfection in the matters of scent, speed, and endurance as can well be, and this, be it noted, is the result of less than 200 years of artificial selection.

The Quorn Hunt—descendants of Meynell's original pack—for long had kennels both at Quorndon and at Billesdon, and has always been the most famous pack of fox-hounds in England, though many other famous packs hunt the county, among which may be mentioned the Fernie, Cottesmore, Belvoir, and Atherstone Hounds. The hunting metropolis is Melton Mowbray, from which the countries of all of the above packs, except the Atherstone, can be easily reached. The circumstances which have contrived to make this district so famous in the hunting world are not very clear, but the explanation is perhaps to be found in the nature of the physical features of the country—the undulating grass lands, with plenty of moderate eminences, but no precipitous descents, and with no impassable rivers or large woods. Before the enclosure of the fields by hedges, which became common some hundred years ago, the pastime must have been enjoyed—jumping apart—under even more favourable conditions than now, and it is said that in those early days an onlooker might often follow the whole of a day's sport from start to finish with scarcely a single change of his view point.

9. Climate and Rainfall.

The chief factors governing the climate of a given locality are the latitude, the direction of the prevailing winds, the proximity of the sea, and the height above sea-level.

The difference of latitude between the north and the south of England is not sufficient to cause very much difference in climate, the average July temperature on the isothermal line, passing through Newcastle-on-Tyne, being only about 4° Fahrenheit less than on that passing through London; but England has a much more equable climate than its latitude alone would lead one to expect. Its latitude is roughly that of Labrador, but its average annual temperature is more nearly equal to that of New York, which lies upon about the same latitude as Naples. For this we have the warm sea current known as the Gulf Stream to thank. Of the other factors mentioned above, however, the winds and proximity of the sea have much greater influence.

The prevailing winds in Leicestershire, as in most parts of the country, blow from the west and south-west. Observations show that the wind blows from this general direction on about as many days as from all other quarters taken together, if we leave out of account the days on which no wind is recorded. Only in the late winter weather, especially in March, when easterly winds are common, is there much departure from the above rule, and of course our county is quite distant enough from the coast for the effect of the diurnal variations—sea breeze and land breeze—to escape notice. Its altitude is nowhere great, and it has no great flat stretches, such as are found in some of the neighbouring counties, across which the east winds seem to sweep with increased fury. We should expect therefore that its climate would be, in most respects, a typical midland climate, with no particular extremes or variations from the average of the surrounding districts, an expectation which is borne out by actual observations.

The effect of distance from the seaboard is noticeable

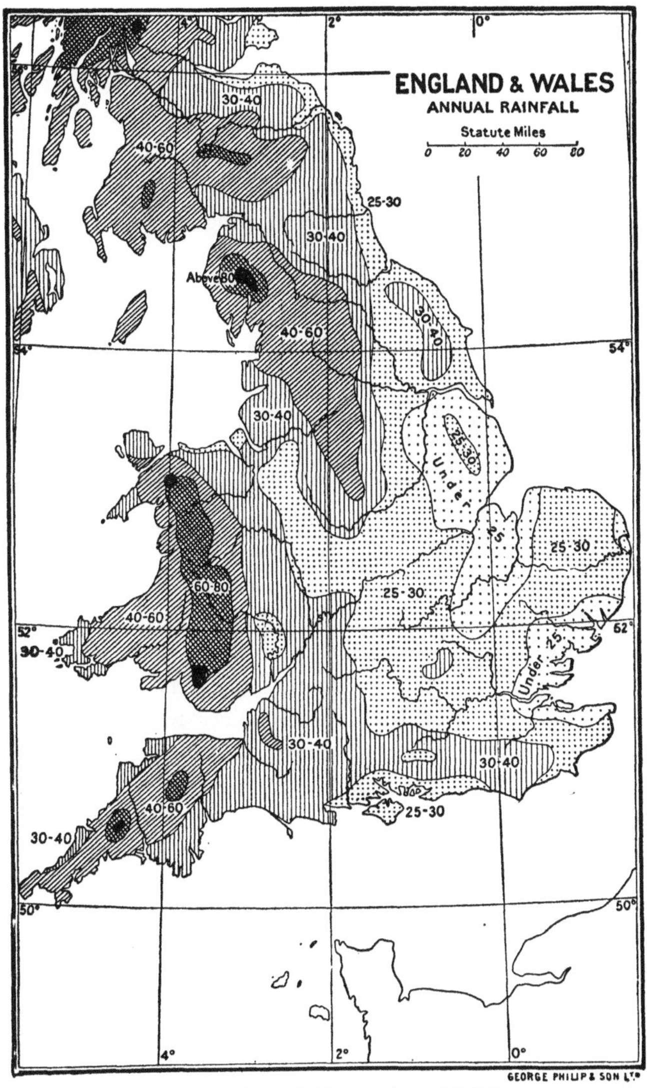

Rainfall Map of England and Wales

in the rather considerable range occurring between the day and night temperatures. Thus, in warm summer weather, a typical variation of temperature might be from a maximum of say 80° Fahrenheit to a night minimum of perhaps 55° Fahrenheit, which is altogether in excess of what we should probably find at the same date recorded by say one of the Cornish stations, where the sea, owing to its great heat capacity, tends to equalise the day and night temperatures of the land. For the same reason the maximum summer temperatures in Leicestershire are often much in excess of those recorded on the south coast. In this respect therefore, the climate of our county approaches what is sometimes referred to as the "Continental" type, but of course the whole of Great Britain only affords a very restricted area as compared with the vast stretches of land of Central Europe and Asia.

Since the prevailing winds over England are from the west and south-west, we should expect that a centrally situated county such as Leicestershire would show an average rainfall less than that of Wales, and more than that of Lincolnshire or the other counties to the east. That such is the case is clearly shown by an examination of the larger rainfall map here reproduced. This map has been prepared from observations taken at a large number of stations, and extending over many years. From a glance at this map the reader can see at once that the Leicestershire average rainfall is between 25 and 30 inches, but from the general form of the 30-inch line, which runs down as a sort of promontory from the north-west towards our county, it might be expected that a more detailed map would show something interesting in this locality. For such a map—here shown—we are indebted to the courtesy of Dr H. R. Mill, late Director of the British Rainfall Organisation. The light dotted

Leicestershire Rainfall

(The light dotted line shows the county boundary)

line indicates the county boundary, and the others are three of the lines of equal rainfall, with their courses traced in more detail than is done in the general map of England and Wales. The map shows that the western side of the county is somewhat wetter than the eastern, and that the Charnwood Forest district, which as we have seen in earlier chapters is abnormal in so many other ways, is the same in respect of its rainfall. The enclosed area in the north-west of our county coincides very closely with the Charnwood district, and shows that the rainfall there is in excess of 27.5 inches, and therefore greater than anywhere else in the county.

As a result of continuous observations, extending over a period of thirty-four years, it appears that the wettest months at Thornton Reservoir, on the edge of the Charnwood district, are July, August, and October, while the driest are February and March. Heavy summer and autumn rains are indeed characteristic of most parts of the Midlands.

The climate of a given district, however, cannot entirely be judged from a knowledge of the rainfall, nor even of rainfall combined with temperature and wind records. Another important factor is to be found in the humidity of the air. Our general health and feelings are perhaps more dependent on this factor than is commonly recognised, or rather upon this factor taken in conjunction with the prevailing air movements. The humidity of the air often differs greatly on a given day in two places comparatively near together. In the Soar Valley the climate is often very relaxing, while on the higher ground, a few miles away, this is not the case. The short railway journey from Leicester to, say, Loseby or Tilton, on the Great Northern Railway, on a cold, damp day in the late winter, is sometimes accompanied by an improvement in the climate which one would have scarcely believed possible.

Lastly, there is still another series of records taken at

most of the meteorological observation stations in the country, namely the duration of sunshine. Maps, showing lines of equal sunshine, similar to those considered for the rainfall, have been drawn, but they are perhaps less instructive. The main results noticed are that, as we proceed northward, the total average duration of sunshine, reckoned for the whole year, gradually decreases. Several places on the south coast receive more than 1600 hours, and nearly all more than 1500 hours in an average year, while to the north of the latitude of about Sheffield, the total is below 1300 hours. One reason, of course, is that as we go north, the elevation of the sun above the horizon is on the whole less than is the case further south, and therefore the sun is more likely to be obscured by clouds. Places on the sea coast also in general receive more sunshine than those at the same latitude further inland. Leicestershire has more sunshine than most parts of Yorkshire or Scotland, but less than either the southern counties, or seaboard counties, such as Norfolk, which are on the same latitude. The slope of the land obviously affects the duration of sunshine in the mornings and evenings, but direction of slope in Leicestershire is about equally distributed between the different points of the compass, and hence no general conclusions as to this can be drawn.

10. People—Early Inhabitants, Place-names, Population.

The earliest people of whom we have any clear knowledge as having lived in this part of England, are those of the Neolithic or New Stone Age. In the south, nearer to the European mainland, a much earlier race of men, those of the Old Stone Age, had lived long before, but no remains of these Paleolithic peoples have yet been discovered in Leicestershire. Many centuries before the Christian era the Neolithic race had given way to a Celtic race, consisting of men who, in many characteristics, doubtless closely resembled their descendants, the modern North Welsh. These were the people in the land at the coming of the Romans.

The Celts, who inhabited what is now Leicestershire, belonged to a tribe known to the Romans as the Coritani. Leicester, or as it was then called Rhagé, was probably one of their principal encampments, but very few of their place-names have survived in a recognisable form, except in the case of some of our rivers and streams, *e.g.* the Avon (Mod. Welsh *Afon* = a river), the Eye, which occurs twice and is evidently related to the Welsh *wy*, meaning water, and the first part of the town-name Loughborough. The termination *dun* (a fort) has also survived in the names of the ancient hill-fortresses, Bardon and Breedon. The country at this early date was very sparsely inhabited, and largely covered by woods and forests.

The Romans, during their 400 years occupation of the land, mixed to some extent with the early British or Celtic tribes, and gave their own names to their principal roads and stations in the district. Thus we have the Fosse Way, so called from the entrenchments running by it, and another trace of their occupation may perhaps be seen in the Strettons, places on the Roman *Strata* or "Street" from Colchester; but, on the whole, as in the case of the Celtic, very few of the Roman names have survived.

Very different is it, however, when we consider the next people who came to dwell in Britain. These were the Jutes, Saxons, and Angles, the latter a warlike people from what is now north-western Germany, who came soon after the departure of the Romans. The Angles drove the Romano-British peoples of Central Britain far to the west, or enslaved them, and settled themselves and their own language firmly in the land. Their place-names ending in *ham* (home) and *ton* (town), the latter after *s* often corrupted into "stone," are found everywhere; their *ing*, denoting "the family of" (as in Skeffington) is quite common; and the other chief Anglo-Saxon affixes such as "*ford*" (*e.g.* Desford), "*stock*," meaning a stockade (Ibstock), "*borough*," a city (Narborough), "*cote*," an enclosure (Huncote), or "*worth*," a farm (Bagworth), are by no means infrequent.

In many parts of England the people of to-day are of fairly pure Anglo-Saxon origin, but in Leicestershire, as in some of the eastern counties, we have to consider still another important addition to the stock, for, on the evidence both of philology and history, Leicester was an important centre of Danish influence. Before the coming of the Danes, in the days of the Saxon heptarchy, England was for long divided into separate kingdoms, one of the largest and most important of which was

Central England before the Coming of the Danes

Mercia, and although the boundaries of these kingdoms were continually undergoing alteration, yet the general position of this central kingdom of the Angles was much as indicated in the map here given. The Danes began their inroads in the year A.D. 830, captured Leicester in 874, but were finally checked by Alfred a few years later.

Alfred, however, was not strong enough completely to crush them, and consequently in 878, at the Treaty of Chippenham (or Wedmore), agreed that the kingdom should be divided between the English and the Danes.

Map of Leicestershire, showing Positions of Place-names ending in *-by*

The line of division crossed Mercia from north-west to south-east, leaving our county on the Danish side; in fact, Leicester soon became one of the five chief Danish boroughs. It is not surprising, therefore, to find that the Danes have left in our county many signs of their occupation. Their terminal *by* in place-names (the Danish *by* is practically the equivalent of the Anglo-Saxon terminal *ton*) is probably commoner here than in any other part of England. The map showing the distribution of these places is also of interest as indicating the decided preference of these people for the river valleys. We know that Sven, at a somewhat later date, sailed right up the Trent into the heart of the Midlands, and our map appears to indicate that his predecessors reached Leicester from Lincoln and the east, by way of the valley of the Wreak, which they probably descended in their long boats.

The Danes, however, did not entirely displace the earlier inhabitants, but in course of time the two races became so intermingled that it is now impossible to say to what extent our ancestors were of Danish or of pure English origin. This fusion is the less surprising when we remember that the two races had much in common. Both were members of the great Teutonic group of races, the fair and blue-eyed type was predominant in both, and their languages had many similarities.

The only other important influx of population into Leicestershire was that of the Normans at the time of the Conquest, but they did not come in sufficient numbers to influence very greatly the language of the common people in central England, although among the ruling classes French was the common tongue until about the year 1360.

The native population of Leicestershire is therefore in the main of Anglian or perhaps rather of Anglo-Danish origin, and this being also the source from which the modern English language is descended, we do not find any marked dialects in the county. Many examples of local pronunciations of common words, as "ship" for sheep, or "tray" for tree, might be named, but there is nothing of the nature of a special dialect.

Passing to the actual figures of population, we find that at the time of the Domesday survey (about 1086),

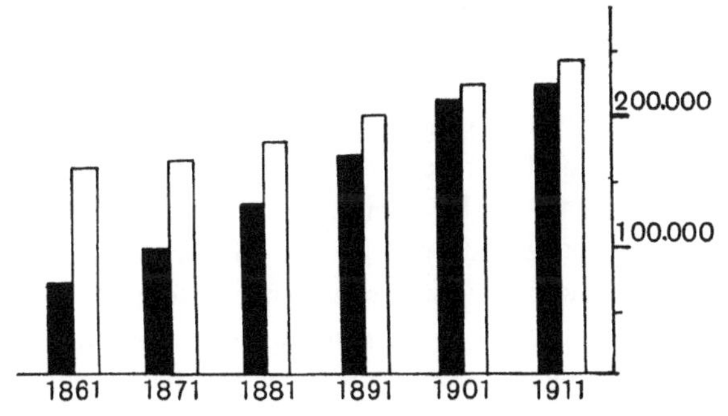

Growth in Population of Leicester (black) and of rest of County (white) during past 50 years

the town of Leicester contained 64 burgesses, 322 houses, and 6 churches, and it is estimated that the whole county had a population of about 34,000. The relatively large number of churches and houses seems to indicate that the wars of the previous century had caused a serious diminution in the population of the town. At the close of the fourteenth century Leicester probably had a population of about 6000, and at the beginning of the nineteenth century, when the first real census returns were taken, it had 23,146 inhabitants, the population of the entire county numbering 155,100. In 1911 the population of the town had increased to 227,142, while that of the county had become 481,115. Thus the population of Leicester, which has increased almost ten-fold during the past 110 years, had only increased about four-fold during the whole of the preceding four centuries.

The rapid increase in the population of Leicester and of the smaller towns, such as Loughborough and Hinckley, during what may be called the industrial epoch, is also of interest in that it took place to some extent at the expense of the rural districts. During the decade ending 1871, when the population of Leicester was increasing by leaps and bounds, that of the purely rural districts showed a very considerable total decrease. The latest census figures, however (1911), happily seem to indicate that this tendency to rural depopulation has been checked.

11. Agriculture

A hundred years ago the Leicestershire population was almost entirely supported by agriculture, or by the accompanying wool-working industries in the county town. Although this is no longer the case, Leicestershire may still be described as being in normal times an important grazing county, but at the same time one in which a considerable amount of wheat, oats, and barley is grown. Although the quantity of human food obtainable from arable land is far greater than from the same area under pasturage, yet the tendency here, as throughout the whole of England, for a long period has been to let the land revert to grass. A Leicestershire writer of some seventy years ago, commenting on this, attributed the change to the enclosure of the open fields, which took place in most parts of the county more than

The Cornlands of North Leicestershire

a century ago, and also to the greater profits accruing from stock-farming with the new local breeds. A more important cause operating throughout the whole country, is undoubtedly to be found in the increasing importation of cheap foreign wheat, which has led to a fall in the price of this cereal, with various fluctuations from 1812, when during the Napoleonic Wars it was as high as 126s. per quarter, to 1894, when it was as low as 22s. 10d., near to which price it remained until the outbreak of the great European War. The Corn Production Act of 1917, the sole object of which was to secure a larger area of cultivation by fixing the price at 55s., did good for a time, but a still higher figure, 100s., has since become necessary. A grazing farm can undoubtedly be conducted with much less outlay and daily attention, and with far fewer labourers, than one under tillage; so that, unless the depopulation of the rural districts during the past few decades can be in some way permanently checked, it is only to be expected that the old tendency to go from tillage to pasturage will continue to operate,

Evidence of the more extensive cultivation of the land in times past may be found in the number of green fields which show "plough ridges." These ridges, which, of course, are to be seen in most parts of the country, are especially common in Leicestershire, even in the eastern portion of the county between the Wreak and the Welland, in parts where, at the present day, the land is almost entirely given up to grazing, and a cornfield is an uncommon sight.

The relative areas of the county in 1918 devoted to permanent pasture, corn crops, roots, and rotation grasses such as clover, and woodlands, are shown in the diagram at the end of the book, and are compared also with the area of what may be called non-agricultural land. The latter includes building land, road and water surface, land occupied by mining, quarrying, and other industrial occupations, and waste heath and moorland. The proportion of the whole area under corn crops is considerably less than that of some of the counties to the north-east, east, and south (Nottinghamshire, Lincolnshire, Rutland, and Northamptonshire), and rather less

Typical Grazing Country of East Leicestershire

than that of the average for the whole country. The soils, however, in many parts of the country—*e.g.* in the vicinity of Loughborough—are excellent for corn crops. Of the land given up to permanent pasture one-third, or thereabouts, is set down for hay, the rich soils in the valleys of the Soar and its tributaries being very suitable for this crop, while the undulating upland country, which makes up the remaining two-thirds, is especially fitted for grazing purposes, both on account of the rapidity with which the rains run from its hill-sides, and of the abundant shelter from the cold winds which it affords.

Of the corn crops, wheat, which grows best on the rather heavy soils of the boulder clay, takes the first place, and constitutes rather less than 50 per cent. of the whole. The other corn crops are oats, barley, and rye. A large amount of oats is grown, a considerable amount of barley, but very little rye (the actual proportions are shown in the diagram on p. 163). Barley thrives best in a warm summer, and on a lighter soil than wheat, while oats prefer a cool and moist climate. Rye, the principal cereal in Northern Europe, is generally grown as an autumn crop in this country, when it provides useful spring fodder for the animals. Beans, although not strictly speaking a corn crop, constitute an important item in the county, chiefly on account of their great value as fodder. There are on the average some two to three thousand acres under this crop.

The root-crops also—turnips, swedes, and mangolds—are grown mainly as winter food for the stock, and the area under potatoes has been increased very considerably in recent years, both by large scale cultivation in fields, and in the very numerous allotments which have sprung up in all urban districts. The amount of fallow land in our county has decreased from being quite a considerable fraction of the total agricultural land some thirty or forty years ago, to a few thousand acres only at the present day.

Turning next to the question of live stock, Leicestershire, in the words of an old writer, can truly "claim to be the cradle and nursery of some of the great early

Leicestershire Sheep

(Some of the few pure-bred "Leicesters" remaining. The property of Mrs Perry Herrick)

improvements in connection with the breed of animals." These were largely originated some eighty or ninety years ago by a certain Robert Bakewell, a farmer of Dishley, near Loughborough. His new Leicestershire sheep was an improved and considerably altered variety of the Lincolnshire breed, and although this has now almost died out as a pure breed, yet the strain remains, and the present Leicestershire sheep is famous throughout the country for its length of wool and other qualities. The Dishley breeds of cattle were also at one time famous, and as a cattle-raising county Leicestershire ranks very high in proportion to its area among the English counties.

Dairy-farming is carried on in almost all parts of Leicestershire, and most of the milk finds its way into Leicester and the other towns, but not a little is sent to the metropolis by way of the two main railway routes which pass through the centre of the county. The County Councils of Nottinghamshire, Derbyshire, Leicestershire, and part of Lincolnshire maintain an important Agricultural and Dairy College at Kingston-on-Soar, where instruction is given in all branches of agriculture.

Cheese-making has for long been an important rural industry, though perhaps now on the wane. Until lately two annual cheese fairs were held regularly at Leicester. In addition to the justly famous red Leicestershire cheese, the county can also claim to be the home of the Stilton. This cheese was first made at Withcote, near the eastern border of the county, but the makers were under an agreement to sell all that they could produce to an innkeeper at Stilton, 30 miles away in Huntingdonshire, an important halt of the coaches on the Great North Road. The fame of the new cheese rapidly spread, and the consumers naturally referred to it as Stilton cheese.

In the seventeenth century Leicestershire was renowned for its pigs, and a writer of that time attributes the superiority of these to the great quantity of beans grown in the district. Melton Mowbray is still famous for its pork pies. At the present time the number of pigs raised is considerably below the average for the whole of the country, while the number of horses is slightly above the average, allowance in each case being made for the size of the county.

Lastly, the woodlands, which comprise some 14,000 acres only, must be mentioned. There are no very large woods, but a great number of small plantations and coverts; the hedge-side timber also, as may be seen in many of our illustrations, is particularly well developed. The ash and the elm are probably the commonest trees, and many a fine ash tree has recently disappeared from the landscape in order to meet the needs of the aeroplane manufacturer.

12. Industries

The two important occupations of agriculture and mining being treated elsewhere, a few of the lesser industries, which might perhaps come rather under the head of manufactures, may here be mentioned. These are, for the most part, carried on in the towns, though for various reasons some have found homes in many of the neighbouring villages.

By far the most important manufacturing town in Leicestershire is the county town itself, which, since about 1840, has risen from a position of comparative insignificance to that of one of the chief manufacturing towns in England. The staple manufactures of Leicester itself at the present time are hosiery, boots and shoes, and elastic fabrics, though many other industries, such as engineering, paper-box manufacturing, cigar-making, and the manufacture of lenses and optical instruments, all contribute to a less extent to the commercial importance of the town.

For very many years Leicester, situated in the midst of an important wool-growing district, has been the centre of the hosiery and allied woollen trades. Even in the fourteenth century it was a staple town for the manufacture of wool, and a locally-made russet cloth was exported to other places, but the woollen trade was not permanently established in the town until towards the end of the eighteenth century. The stocking-machine, which had been invented some time before by a Nottinghamshire clergyman, was introduced into Leicester about 1680, and about forty years earlier into Hinckley. In 1792 it was said that "the manufacture of stockings in the town and county of Leicester is the largest in the world." At this date more than 40 per cent. of the entire population of Leicester was engaged in the trade. The number of stocking-frames in the United Kingdom in 1844 is stated to have been 48,482, and of these 20,861 were in Leicestershire. Many of them may still be seen in the villages, though, of course, steam-power and immense works have long ago replaced the single frames. Nowadays the trade is concerned not only with the manufacture of stockings, but with jerseys and all kinds of woven goods and fancy hosiery. Much cotton hosiery

An Old-time Stockinger at Work

is also made at the present time, though originally wool was the only raw material.

As well as the hosiery factories, there are worsted spinning mills. At the largest of these, the crude wool received in bales from abroad goes successively through all the operations of sorting, "willeying" (clearing of refuse), scouring, drying, combing, carding, and spinning into yarn, as well often as dyeing, before being finally manufactured into the various articles of clothing referred to in the last paragraph.

The manufacture of elastic web, which in the days of elastic-sided boots was an important one, has now to some extent declined, but there is still a considerable export trade from the town of elastic-woven goods, and several new mills for the manufacture of plain india-rubber goods have recently been established in the county at Market Harborough.

At the present time, however, Leicester owes its prosperity even more largely to the manufacture of boots and shoes than to the woollen and hosiery trades; at any rate the number of operatives employed in the leather trades—some 40,000 or more—is larger than in the older woollen manufactures. The tanning of leather and the manufacture of boots and shoes have been carried on in this country for some hundreds of years, but the industry was revolutionised in about the year 1860 by the introduction of labour-saving machinery. During the serious labour dispute which attended the introduction of machinery at Northampton, then the chief home of the trade, the industry became established in Leicester, and

A Modern Hosiery Factory

now Northampton and Leicester are its recognised centres, and to some extent vie one with the other for the premier position. But the making of boots and shoes in the two counties is not by any means confined to the large towns: in Hinckley especially, and certain of the neighbouring villages, the manufacture seems to flourish as well as in Leicester itself, as also in the large villages of Wigston, Syston, and Anstey, all within a few miles of Leicester. What is said to be the largest boot and shoe factory in the world is situated in Leicester, but before the war the town was beginning to feel the effect of increasing American competition. During the war millions of pairs of army boots were manufactured, both for our own armies and for those of our Allies. Several of the large firms specialise in women's and children's boots, while others are engaged in the manufacture of canvas and rubber shoes, or of laces, linings, polishes, and other accessories. Moreover, an important branch of the engineering trade has become established in the town to supply the great demand for boot and shoe machinery. Some of this is of a highly complex and delicate nature, for example the automatic "edge-setting machine," which is worked by oil pressure, can deal with over 800 pairs of boots or shoes per day, and is considered one of the most wonderful machines used in the boot trade.

At Loughborough, which is the second most important manufacturing town, lace-making was at one time a considerable industry, and now the finer branches of the hosiery trade are largely carried on. Here, too, is a famous bell foundry, where Great Paul was cast in 1881.

But of recent years Loughborough has become more and more an engineering town. The Brush Electrical Engineering Works covers an immense area, and electric traction and power plant and apparatus are exported to all parts of the world by this firm.

Hinckley, the third town in the county, in addition to

Paving Stones : Croft Granite Co.

its boots and shoes and hosiery, makes a special type of needle which is of very great importance in the hosiery trade.

Other industries within the county which have more than a local importance include the manufacture of glazed sewer pipes, sinks, and all kinds of sanitary apparatus, which is carried on in parts of the populous district stretching over the north-west boundary into Derbyshire. A firm having works at Overseal in this district undertook and carried to completion a few years ago a scheme for the complete drainage of the city of Rio de Janeiro. Another large local industry, of somewhat similar nature, is that carried on by the company which works the granite quarries at Croft, between Leicester and Hinckley. Here immense quantities of paving slabs are manufactured of artificial stone, which is made from granite chippings and slow-setting Portland cement. The product, which is known in the trade as Croft Adamant, is more resistant than the natural granite.

An important branch of the dyeing and cleaning trades, which undertakes work from all parts of the country, has been established near Hinckley, and another minor industry which must complete the list is the manufacture of wicker chairs and baskets. Osiers for this purpose are extensively grown in the Soar Valley a few miles to the north of Leicester, and some large firms are engaged in the wicker chair trade. From the locally grown osiers immense numbers of baskets for the transport of shells were made during the war. Another firm specialises in the manufacture of very high-class cane chairs ; these, however, are made from imported canes.

13. Mines and Quarries

The most important mineral obtained in Leicestershire is undoubtedly coal, in spite of the fact that, compared with some of the coal-fields of Great Britain, that of Leicestershire is small, occupying an area of not more than about 20 square miles. It is, nevertheless, an extremely busy area, especially in the neighbourhood

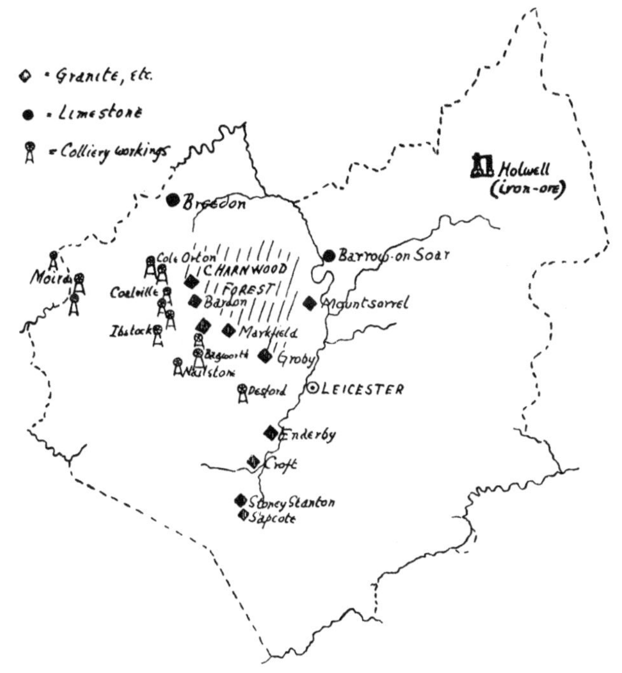

Leicestershire Mines and Quarries

of Coalville, and it is of great importance as being one of the nearest coal-fields to the metropolis, whither much of the coal proceeds by way of the North-Western main line.

In times past the coal-field was probably of much

greater extent, and its geology is of great interest, owing to the close proximity of the volcanic rocks of the Charnwood district. At Whitwick, for instance, it was found that a bed of volcanic dolerite had burnt to cinders much of the coal on which it was resting, giving clear evidence of the late intrusion of the volcanic rock. Many "faults" also are found, one of which, running from Bardon Hill to Ticknall, and forming the eastern boundary of the coal-field, has displaced the strata through no less than 2000 feet.

As will be seen from the map annexed, the coal-field is divided into two separate productive areas, that around Moira on the west, which extends into Derbyshire, and that near Coalville more to the east. During recent years coal has been worked as far south as Desford, a point only 8 miles from Leicester. A typical coal boring on either side passes through the overlying rocks of the Trias before reaching the coal measures, which consist of alternating beds of sandstones and shales, with numerous seams of coal on different horizons. The most important or "main" seam in the western coal-field is 14 feet thick, and ten or more other seams, varying from 3 to 7 feet in thickness, are also met with before reaching the millstone grit and underlying non-productive Carboniferous rocks. On the eastern side even more coal seams are passed through, but none of greater thickness than about 8 feet. The seams of the two divisions have not been correlated, and in many of the workings a given seam is often found to come to a sudden stop owing to the existence of one or other of the numerous "faults" already alluded to. Abortive borings have shown that the coal measures do not extend much to the south. At Elmsthorpe, not far south of Desford, a boring passed for 540 feet through the Keuper marl, and then entered Cambrian rocks, and the same kind of result was obtained at numerous other places. In the neighbourhood of Moira valuable beds of fire-clay are found associated with some of the coal seams, and this clay is extensively made into fire-bricks and muffles. Other clays in the same neighbourhood are utilised in the manufacture of the drain pipes, etc. already mentioned.

Next in importance to the coal among the minerals of the county come the igneous rocks, and the value of these is also to a great extent due to the fact that they are among the nearest of their kind to London and the south. The most valuable of these is granite, together with the closely allied rocks, diorite and syenite, which differ chiefly from granite in having hornblende instead of quartz as one of their chief constituents. Leicestershire granites are well known in most parts of the country, and are used extensively for road-making in southern and central England. These igneous rocks, as previously stated, make up the chief part of the Charnwood Forest district, but the points at which they are worked often lie at considerable distances to the south as well as on each side of the Forest area. This is seen in the map on page 73, which also shows that the coal measures come to a sudden stop where the igneous rocks come to the surface.

A typical granite quarry (really syenite) is that at

Croft, a view of which is here given. The stone is very tough and durable, and is worked into kerbs, masonry, cubes, etc., as well as being made into the artificial stone already described. This quarry alone yields some 80,000 tons of stone per annum, and the total yearly value of

The Croft Granite Quarry

the whole of the igneous rock produced in the county amounts to about £350,000. Of the true granite (*i.e.* conglomerates of felspar, quartz, and mica) quarried in the county, that from Mountsorrel, which has a pink or grey colour, is famous throughout the world. The large isolated mass of granite at this place, forming a noticeable eminence overlooking the Soar Valley, has now to a great extent been cut away.

At one time slate also was quarried in the neighbourhood of Swithland, but the slate quarry there is now disused and filled with water. The Leicestershire slates are very durable, but thicker and consequently heavier than the Welsh slates which have displaced them.

Disused Slate Quarry at Swithland

Thick slabs of slate were until recently used by the local farmers for cheese-presses.

Valuable limestone quarries exist at one or two places in the county: at Breedon and Breedon Cloud in the north-west the fossiliferous magnesian limestone or dolomite belonging to the Carboniferous System is worked, and at Barrow-on-Soar a more important limestone of argillaceous nature is quarried in the Lower Lias rocks. The quarries here are said to be the finest in England for the production of hydraulic lime and cements, and they are also famous for their fossil remains. There are also oolite quarries near to Stonesby, in the north-east, from which a valuable building stone is obtained; and gypsum, a mineral of some importance in the neighbouring county of Nottinghamshire, was at one time worked near to Loughborough.

Lastly, a certain amount of ironstone is found within the county, though it is only worked at all extensively in one district, viz., in the neighbourhood of Wartnaby and Holwell, a few miles to the north of Melton Mowbray. The rock bed here, belonging to the Middle Lias, is very ferruginous in places, and the soils of the many escarpments of this rock throughout the district have generally a fine red colour. The illustration on the opposite page shows an open iron-working, where the covering masses of boulder clay can be seen.

14. History

In the brief sketch of the history of Leicestershire, which is all that can be given here, we shall begin with the Norman period; not that the recorded history dates from no more distant times, but because the earlier periods have already been to some extent dealt with in previous chapters, and other references to Pre-Norman times will be made in later pages.

That Leicester was then a town of some importance

Quarrying Iron Ores

is clear from the fact that at the Norman Conquest, it possessed six churches. Four of these (though not retaining much of the original structures) still survive as the principal parish churches of the town (see p. 87), but the bishopric which existed in Anglo-Saxon times had been removed in 874 to Dorchester, and afterwards to Lincoln. There was also a Royal Mint in the town, which continued to issue coin of the realm from the reign of the Saxon King, Athelstane, until that of Henry II.

The Norman period begins with the capture of the town in 1068 by William the Conqueror, who had a few months previously subdued the neighbouring town of Warwick. We owe our knowledge of the actual conditions existing at this time chiefly to the wonderful record of the survey of the whole country compiled by William, and known as Domesday Book, which, preserved in manuscript at Westminster through the intervening centuries, was first printed in 1783. From it we learn that the county of Ledecestrecire was divided into four "wapentakes," three of which met at Leicester, but the town itself was a "free borough" standing on no man's land. We also learn that the principal landowner was one Hugh de Grentesmaisnell, who was Governor of Leicester, and had fortified strongholds both in that town and at Hinckley. Other important land-owners were the King himself, the Archbishop of York, and the Bishop of Lincoln. Hugh appears to have been a harsh ruler, but neither he nor his son Ito oppressed the people for long, for in A.D. 1107 the estates came into the possession of Robert de Beaumont. Robert, who was a great upholder and friend of Henry I, by whom he was created Earl of Leicester, was not only a wise ruler of the town, but became one of the most powerful nobles in the land. He was the first of a long line of Norman earls of Leicester, and ancestor of Simon de Montfort (killed A.D. 1265), in whom the line terminated. These Norman earls were not the first Earls of Leicester, but they quite supplanted the older line of Saxon earls, just as their castle at Leicester was built upon the site of the older Saxon castle which had been destroyed in 1068. It was during this period and that immediately following—say from 1107 to 1399—that Leicester reached its point of maximum importance in medieval history, and from what is known of the many important events there, its castle must often have been the scene of much magnificence and pomp. Of the castle itself little but the site now remains, but the church of St Mary de Castro, now the parish church of St Mary, still shows much early Norman work (see p. 82).

In 1173 Leicester was destroyed by Henry II, and for twenty years or more appears to have lain in utter desolation. In 1201 a great meeting of the barons was held at the castle to discuss their grievances against King John, and in 1218 Simon de Montfort, then only eight years old, succeeded to the earldom. Simon, though at first enjoying the favour of the King (Henry III), and indeed married to his sister, eventually became more and more the champion of the people. He entertained the King and Prince Edward at Leicester Castle on at least one occasion, but shortly afterwards, the

King having been defeated by the barons at Lewes, Simon summoned what was the first really representative Parliament of the people. He was able to do this by virtue of his powers as Lord High Steward of England, an office which he held by hereditary right as Earl of

Norman Sedilia : St Mary's Church, Leicester

Leicester. In local affairs Simon ruled as wisely as in those of the nation, and among other generous acts may be named his gift to the burgesses of the town of a large area of land on the southern side of Leicester, which includes what is now the Victoria Park.

In 1265 Henry III conferred the confiscated titles of the great earl upon the Earl of Lancaster, who thenceforth became first Earl of Lancaster and Leicester. From him was descended a line of Lancastrian earls which lasted until 1399, when, in the person of Henry Bolingbroke, the Earldom of Leicester, as well as the Dukedom (as it had now become) of Lancaster, became absorbed in the Crown of England. The earls of more recent date, some of whom, such as Robert Dudley, became famous in their day, had no real connection with the town.

The last and most famous earl of the Lancastrian line was John of Gaunt, father of Bolingbroke, who seems to have spent much time at his castle of Leicester, and probably often visited his smaller castle of Earl Shilton, near to Hinckley, a popular hunting-seat of the later earls. It is interesting to read in his Will, written at Leicester, that after various bequests to the King, Richard II, to his wife, Catherine Swinford, and to his son, soon to become Henry IV, he bequeathed his garment of red velvet embroidered with gold suns to the "new Church of Our Lady." This church was not the St Mary de Castro already mentioned, but the collegiate church of St Mary in the Newarke, which adjoined the castle proper on the south side. It is said to have been a magnificent building, though small, but unfortunately hardly a trace of it now remains.

In the history of Leicestershire, John of Gaunt is perhaps most worthy of note in that he gave his valuable patronage to one whose life work was destined to exert far greater influence on posterity—John Wycliffe. Wycliffe, who for a great part of his life dwelt at the

Rectory of Lutterworth, doubtless often preached at the Castle Church of Leicester, denouncing the greed and ostentation of prelates and Pope. These doctrines were undoubtedly most acceptable to the Duke, and on the other hand, it is very doubtful whether the preacher would have survived the Papal attacks made upon him without the support of his great patron. Hence we find linked together with mutual benefits in those distant times both the powers of State and the "Morning Star of the Reformation," and both, moreover, closely associated with our town and county.

After the death of John of Gaunt, the importance of Leicester seems to have declined. On several occasions, however, during the next half century, Parliament met in the town—in 1414, when the "Fire and Faggot Parliament" attempted to exterminate the heresies spread by Wycliffe's teaching; in 1426, when the young Henry VI knelt before the Altar of St Mary's to be received into the ranks of knighthood; and again in 1450, a few years before the outbreak of the Wars of the Roses.

When Richard III passed through the town on his way to the fatal Bosworth Field (1485) the castle seems to have so far lost its former grandeur as to be deemed unsuitable to receive him, for we learn that he was lodged at the Blue Boar Inn, a fine old building which survived until 1838. Of Bosworth Field itself much might be written, but being of national rather than local interest, it need not detain us here. After the defeat and death of Richard, his mutilated body was conveyed again to

Leicester, twelve miles distant, tied upon a pack-horse. His first burial-place is uncertain, but it is supposed that his bones were eventually flung into the river, near the Bow Bridge.

During the Tudor period the most important events of

The Blue Boar Inn, Leicester
(From an old print)

local history centre round the pathetic death, in 1530, of the great Cardinal Wolsey at Leicester Abbey, shortly before its dissolution, and secondly round the even more pathetic figure of Lady Jane Grey. Bradgate Hall, at which place she was born and spent her childhood, was built at the beginning of the sixteenth century by Thomas, Lord Grey, and remained the chief seat of the family until it was destroyed in 1694. Bradgate was a "parcel" of the Manor of Groby, where long before had been a Norman castle, and where had lived for some years Elizabeth Woodville, Queen of Edward IV. Lady Jane Grey was married in 1553, while quite a child, to Lord Guildford Dudley, son of the Duke of Northumberland, and a few months later was proclaimed Queen of England, quite against her own wish. She was beheaded in February 1554, together with her husband.

In the Civil War Leicester sustained a short, but rather important siege, and the county was the scene of several minor engagements. The chief party leaders on the King's side were Colonel Hastings, his father the Earl of Huntingdon, living at Ashby-de-la-Zouch, and William Cavendish, Earl of Devon, who lived at the Abbey Mansion; while the many sympathisers with the Parliament included the Earl of Stamford, his son Lord Grey of Groby, Sir Arthur Hesilrige, and Mr Archdale Palmer, the High Sheriff of the county. Immediately before the outbreak of the war Charles was twice in Leicester, but seems to have had a rather cold reception, and on his finally raising his standard at Nottingham in August 1642, hostilities at once commenced in our neighbourhood. Prince Rupert, after attacking the Earl of Stamford's house at Bradgate, withdrew the Royalist forces to Queniborough, but shortly afterwards left the county, and abandoned Leicester to the Parliamentarians. During the next few years fighting took place at Belvoir Castle, which changed hands several times, at Melton Mowbray, Market Harborough, and in

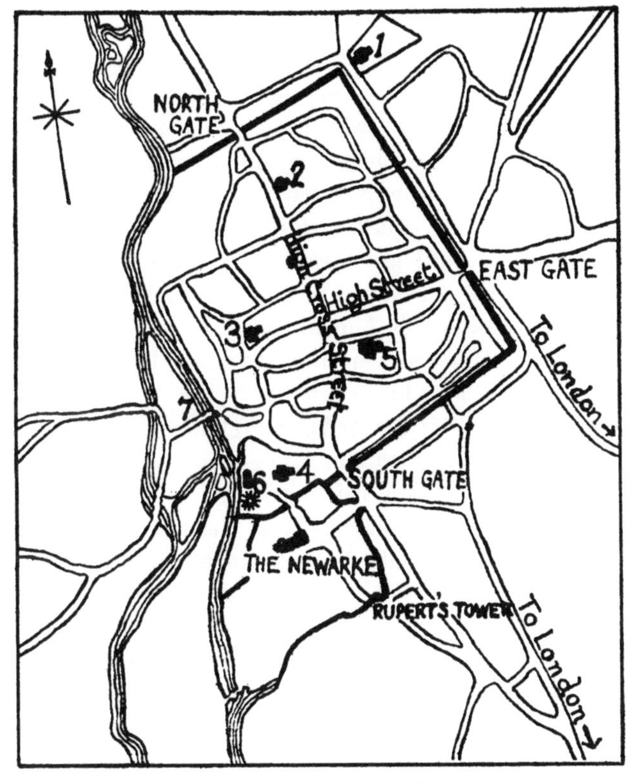

Leicester at the time of the Siege, 1645

1. St Margaret's Church. 2. All Saints' Church. 3. St Nicholas' Church. 4. St. Mary's Church. 5. St Martin's Church. 6. The Castle and Castle Mound. 7. The West Bridge.

the neighbourhood of Ashby-de-la-Zouch, from which place as base Hastings fought a long series of fierce encounters with the Parliamentarians at Leicester. It was not until 1645, however, that any attempt was made by the Royalists to capture Leicester. This Charles eventually decided to do in order to draw off the enemy forces from Oxford, which place Fairfax was then besieging. After a short bombardment the town walls were breached at the Newarke and in other places, the Royalist forces (again under the command of Rupert, though the King himself was also present), fought their way in, and the town was sacked. The fall of the town caused a great stir in London, but the Royalist triumph was only short-lived; in the next month, June 1645, the King's forces were badly defeated at Naseby, just to the south of Leicestershire, and Cromwell, pursuing the King to the north, regained possession of the town without a fight. Belvoir Castle held out for some time longer for the King, and was indeed one of the last fortresses to be taken, while Ashby-de-la-Zouch finally surrendered in March 1646, on the condition that its defenders were allowed to march out with loaded muskets.

From the time of the Commonwealth down to the beginning of the nineteenth century, Leicestershire took but little part in our national history. Towards the end of the eighteenth century, however, the woollen trade became established, and thenceforward the industrial population began to increase, at first slowly, but afterwards by leaps and bounds. The two most interesting dates in the local history of Leicester during this latter period are perhaps 1785, when the first mail coach passed through the town on its journeys between London and the North, and 1794, when the town gates were removed. The gates were the last remnants of the ancient fortifications which ever since Roman times had encircled the town. The walls themselves had for many years previously been in decay and in part removed, but

Tudor Gateway and St Mary's Church, Leicester

after the demolition of the gates, their sites were built upon. The gates now remaining are not the town gates proper, but openings into the Newarke, an adjoining area which was first enclosed by one of the Lancastrian earls.

15. Antiquities

When the Romans first came to Britain, they found the country peopled by a race of men quite different from themselves. How long this race had dwelt in the land it is almost impossible to say, but it is agreed by most that the time from the first coming of man down to the arrival of the Roman legions was immeasurably greater than that which has since elapsed.

The past nineteen centuries or so alone have any claim for treatment in the actual history of this country, so that the study of the records left by man from all the previous centuries is referred to as pre-history. Obviously there is no sharp dividing line between the two periods, for, although we may say that history dates from when men began to leave written records of their doings, yet the first written accounts that we have are so meagre that they can only be regarded as supplementary to the more abundant remains of a non-literary nature. Of the antiquities to which reference is made in this chapter, some belong to the pre-historic period, but others, which are of greater interest in our particular county, date from Roman or later times, and are therefore contemporary with the earliest historical records.

Pre-history is usually grouped into the broad divisions shown in the following table, and some of these are again sub-divided into lesser periods, just as are the geological epochs mentioned earlier in this book; but, like the geologist, the pre-historian is generally very chary of attempting to estimate the actual duration in years or centuries of the given periods.

THE SUCCESSIVE AGES OF MAN IN GREAT BRITAIN

Period.	The Predominant Race in Britain.	Present-Day Descendants and Chief Remains.
Palaeolithic, or Old Stone Age	" Palaeolithic man." Cave dwellers and hunters in the river valleys. Probably red-haired.	Extinct. *Remains:* None in Leicestershire. Roughly chipped flints in southern counties.
Neolithic, or New Stone Age	" Neolithic man," or Iberians. A race of dark, small men with long skulls (dolichocephalic).	Almost extinct; the race perhaps survives in S. Wales and the West of Ireland. *Remains:* Well-worked and often polished stone and flint implements.
Bronze Age (perhaps 1000 B.C.)	The Goidelic Celts. Bigger men with rounder skulls (brachycephalic)	*Descendants:* the Gaels of Scotland and Ireland. *Remains:* Rude pottery and implements of both bronze and stone.
Iron Age	The Brythonic Celts. Men of fair hair and blue eyes.	*Descendants:* Many of the Welsh and some of the western English. *Remains:* Better pottery as well as implements of both bronze and iron.

Up to the present no remains of men of the Palaeolithic Age have been found in Leicestershire. Sir John Evans, one of our greatest authorities on such matters, pointed out some years ago that a line drawn from the Wash to the Severn estuary marks fairly accurately the northern limit to which Palaeolithic man penetrated, but there would seem to be no very obvious reason why he should not have gone further. Of Neolithic man on the other hand several interesting relics have been found, but, as a rule, in positions such that little further

Neolithic Hammer-head

information can be gleaned from their surroundings: hence it often happens that the antiquary has considerable difficulty in deciding to what part of the New Stone Age a given object should be assigned. The several beautifully shaped hammer-heads and "celts," both of flint and stone, now in the Leicester Museum, appear to date from about the end of that period, but it is quite probable that they were in use, or indeed some may have been made, during subsequent ages. From the more complete records found in other parts of England, we know that the men of the New Stone Age were tillers of the soil, that they built their own huts, and wove their own garments. They were, in fact, of quite a different race from their predecessors, whom they seem to have followed only after a considerable gap in time. They gave place more or less imperceptibly to their successors, the men of Celtic race, who first learnt to work in bronze, and—what is perhaps more remarkable—to smelt the ores from which the bronze was obtained.

A good number of relics of the Bronze Age have been discovered at different times. Perhaps the most interesting find in our county of this period was that made in cutting a drive on Beacon Hill in 1858. Two spear-heads, a gouge, an armlet, and a celt, all of Bronze, were found in quite a small area, and make it extremely probable that the ancient encampment, which can still be traced on the hill, was inhabited by men of the Bronze Age. Moreover, a better site could hardly have been found, for from the hill the enemy could be discerned in ample time to admit of a warm reception being prepared for him. Numerous examples of later, but still Pre-Roman date, have also been found in Leicestershire, and some of these are of quite exceptional interest to the antiquary. The iron implements naturally have not survived in very good condition, but the early Britons of those times worked also, like their predecessors, in bronze and in clay. Fine specimens of early British pottery, in the shape of cinerary urns, have been found at Mountsorrel and Syston, and the elaborately-ornamented bucket of wood and bronze illustrated here was also unearthed at Mountsorrel. Very few specimens of this latter type of work are known. Lastly, good examples of ancient earthworks may be seen in many places: that near Ratby has already been mentioned,

Early British Urn

Wood and Bronze Bucket

and others exist at Houghton, Billesdon, Mountsorrel, Hallaton, Tilton, and Castle Donnington. There is much difficulty in assigning even an approximate date to the construction of these earthworks, though the above-mentioned are for the most part undoubtedly of Pre-Roman origin. The great mound on which Belvoir Castle is built has been stated to be the work of early man, but more probably it was a natural hill site, strengthened and adapted for defensive purposes by the still clearly-marked earthworks.

The antiquities of known Roman origin include the roads (described later), a portion of an old wall at Leicester (illustrated on page 96), the sites and remains of several Roman villas in different parts of the county, several finds of Roman glass and pottery articles, various

Roman Pottery found at or near Leicester

hoards of coins, and numerous miscellaneous articles, such as the small bronze steel-yard now in the Leicester Museum, which is to-day still in good enough condition to fulfil the purpose for which it was constructed so many centuries ago.

Of these Roman antiquities, the most important have been found at Leicester itself, which was one of the chief stations on the great road running from Lincoln to the south-west of England when the surrounding district was apparently very sparsely inhabited. Our county town, in the fourth century, was probably enclosed by strong walls on the north, south, east, and west, though according to some authorities the town's only defence on the west was afforded by the river. Of these walls, only one small portion (now known as the Jewry Wall),

The Jewry Wall, Leicester

remains, about 25 yards only in length, but sufficient to exhibit very clearly the Roman method of arch-construction in brickwork. This interesting relic seems to have been part of a western gateway to the town; it was added to at a later date, though during the Roman occupation. It should be remembered, in this connection, that the duration of the Roman occupation of

England was approximately the same as that of the time which has elapsed since Henry VII was on the English throne.

Although the Roman walls have for the most part disappeared, the lines along which they ran are still clearly marked by some of the thoroughfares of the

Roman Pavement discovered at Leicester

modern town, and their positions may be seen clearly in the plan of the town on page 87. In Roman days the principal buildings were situated near the junction of High Cross Street and St Nicholas Street: the former crossed the town from the South Gate to the North Gate, while the latter, together with the modern High Street, ran across from west to east.

An ornamental cross existed at this point until as recently as 1836, and its position is now indicated in the pavement. Near to this point many interesting remains have been found, including the beautiful fragment of tesselated pavement here shown, which still remains *in situ* in the cellar of a shop opposite the churchyard of St Nicholas, and the bases of two massive Doric columns, which were discovered during the rebuilding of the tower of St Martin's church in 1861. A second tesselated pavement was discovered about half-a-mile away, and is now preserved under one of the arches of the Great Central Railway. This also is a beautiful piece of work both in construction and in colour scheme, the latter being a combination of two shades of red brick with blue slate and various shades of limestones.

In 1850 the foundations and floor of an important Roman villa were laid bare at a position some distance to the west of the town, and other mosaics discovered. One of the best portions of these was removed to the town museum, but the rest of the work was unfortunately destroyed. Other Roman villas have been found at Medbourne, lying just off the old Gartree Road which ran from Leicester to the south-east, and at Rothley, and an important find of Roman glass was made at Barrow-on-Soar. But perhaps of still greater interest to the archaeologist is the Roman milestone discovered at Thurmaston on the Fosse Way, two miles, or rather "two thousand (double) paces from Ratae." Its inscription—an elaborate dedication to the Emperor

Hadrian, who had recently passed through the town—has all but perished (p. 125).

Of the centuries which elapsed between the departure of the Romans and the coming of the Normans, the

Bronze Fibulæ (or Clasps) of the Anglo-Saxon Period

archaeologist can tell us less than of the Roman period. Neither the Anglo-Saxon nor the Danes were particularly distinguished as builders, though the former had a keen sense of beauty in the matter of personal adornments. This we learn from the many artistic bronze clasps and brooches which have been found, some of which are illustrated here. The Saxons also probably built the mound on which stood the first Leicester Castle. Other relics of these times include an amber necklace, and various daggers and iron implements, now in the Leicester Museum, a churchyard cross at Rothley, and one or two architectural fragments which belong rather to the next chapter than to this.

16. Architecture—(*a*) Ecclesiastical

Although Leicestershire cannot boast of any important cathedrals or abbeys, yet many of its parish churches, both in town and country districts, present points of great interest to the student of architecture and to the ecclesiologist. Before proceeding to a description of some of the more important of these churches, it will be well perhaps to give a brief general account of the chief architectural styles which have prevailed in this country since men first learned to build in stone and similar materials.

The accompanying table shows what may be called the characteristic national styles: some of these had their counterpart on the Continent, especially in Northern France, where a given style often came to perfection somewhat earlier than in England, but the dates given must only be taken with very great latitude. The Early English, Decorated, and Perpendicular are the so-called Gothic styles.

Style and Approximate Dates.	Leading Characteristics.
Pre-Norman or Saxon. (Up to about A.D. 1060)	Walls largely of rubble: no buttresses: triangular or round-headed windows: square towers, "long and short work" at their corners.
Norman. (A.D. 1060–1150.)	Large semi-circular arches: recessed doorways: simple zig-zag and "dog-tooth" mouldings: massive pillars and columns: square, low towers: flat buttresses.
Early English. (A.D. 1190–1250.)	Long, narrow, lancet-headed windows: vaulted roofs: slender grouped columns: recessed pointed arches, with elaborate mouldings, but much restraint and delicacy in ornament.
Decorated. (A.D. 1290–1360.) The earlier portion also known as geometrical.	Broader windows, much ornamented with tracery: crocketted pinnacles and spires: vaulting ribs of more intricate patterns.
Perpendicular. (A.D. 1360–1530.)	Perpendicular and horizontal lines in window tracery: embattled roofs and towers without spires: elaborate vault traceries, giving rise to "fan vaultings" in Tudor times. Is confined to our own country.
Renaissance, Jacobean, and later Styles.	Introduction of ornament of "classical" styles.

Very few indeed of the Leicestershire churches belong solely to any one period: most have been added to or repaired in later years in the new styles then prevailing. This lack of purity of style, however, in many cases adds greatly to the general interest, while the resulting effect, as may be seen for instance by a glance at the fine church of Melton Mowbray (page 109), is often most agreeable.

The oldest Leicestershire church, and indeed one which claims to be among the oldest in the country, is St

St Nicholas, Leicester

Nicholas', Leicester. Much of the nave is undoubtedly Saxon, and some of the Saxon work, *e.g.* two windows in the nave, is built of Roman tiles which were probably taken from the Roman remains close by; indeed, a temple existed on the very site of the present church, even before the introduction of Christianity into Britain.

Externally St Nicholas' is chiefly Norman, and in the chancel there is a beautiful Early English pillar with detached shafts: the church also has a central tower, which is uncommon in Leicestershire, the only others being St Martin's, Leicester, Melton Mowbray, and five small village churches. Few other churches exhibit

St Margaret's, Leicester

St Peter's, Belgrave

Early English and Norman Arches

any trace of Pre-Norman work. Birstal has a window and font, and Breedon a very fine sculptured frieze, which is believed to be Saxon. At Rothley also, there is a late Saxon churchyard cross.

Examples of Norman construction are fairly numerous, but all the churches built during this period contain later additions. The most beautiful Norman work is perhaps that found in the nave and chancel of St Mary's, Leicester, an example of which is given in the illustration on p. 82. The finest Norman doorway is at Belgrave, a church which also contains some beautiful vaulted sedilia of later date; and a good example of a Norman tower (the lower part perhaps pre-Norman) may be seen at Tugby in East Leicestershire. All the old Leicester churches except St Margaret's exhibit Norman work in one form or other, as do the village churches of Arnesby, Hallaton, Thurlaston, and Twyford.

During the Early English period, there seems to have been no great amount of church-building in the county, and the use of the Norman semicircular arch appears to have persisted here longer than in most parts. There is no typically Early English church, but excellent examples of the work of this period may be seen in the south doorway of St Margaret's, Leicester, in the clerestory of St Mary's, Leicester, at Castle Donnington, and in the village churches of Caldwell, Great Easton, Medbourne, Somerby, Stathern, and Waltham-on-the-Wolds.

Towards the end of the Early English period, on the other hand, church-building in the county began to flourish. The years between about 1250 and 1399, say from Simon de Montfort to John of Gaunt, were, as we have seen, important ones in the secular history of Leicester, and especially during the first part of this period many fine new churches were built, and many of the older ones enlarged and decorated. The style adopted was naturally that of the times, namely, a transition from Early English to Decorated, and examples of churches in the fully developed Decorated style are not at all uncommon. The window tracery of some of the churches of this period is most interesting and varied, the churches at Stoughton, Barkby, Belgrave, Evington,

St Margaret's, Leicester *St Mary's, Stoughton*

Decorated and Perpendicular Windows

Gaddesby, Hallaton, Market Harborough, and Melton Mowbray being all well worth visiting from this point of view alone.

During this period many beautiful spires were erected, and the churches of East Leicestershire compare well in this respect with the better known examples in Northamptonshire and Lincolnshire. The towers and steeples are usually at the west end of the church, and the plan commonly adopted is that of nave with two aisles and aisle-less chancel, though cruciform churches with well-

St Luke's, Gaddesby

developed transepts, as at Medbourne, are not uncommon. The spires are often of the "broach" pattern, *i.e.* grow directly out of the tower, without any parapet on the latter, and the excellent treatment of the belfry stage constitutes quite a local style, which is well seen in the Early English spires at Hallaton, Great Easton, Gaddesby (see p. 106), and Oadby, and in the Decorated spires at Market Harborough and Kirby Bellars. A good modern example of the same style is the spire of St Martin's, Leicester (p. 118). The last-named church has a history going as far back perhaps as that of St Nicholas, Roman foundations having been discovered on the same site. The present church is largely Early English, with later chancel and windows, and the modern spire rises from four lofty, pointed arches where once stood the low Norman arches which supported the old tower.

The Black Death in 1348 greatly checked all building, and it was not till towards the end of the century that much was done, and the Perpendicular style sprang into existence. Perpendicular churches are less common, but much excellent work of this period may be found in the fine church of St Margaret, Leicester, as well as at Frowlesworth, Ashby-Folville, Lutterworth, Loughborough, and Melton Mowbray. The important church at Ashby-de-la-Zouch is almost entirely in this style, as are the tower and remarkably slender spire at Queniborough.

It is a fact worthy of note that in the list of names here given nearly all of the churches mentioned are on the eastern side of the county. This may be an expression of the greater agricultural prosperity of this district in medieval times, but it is also due to the proximity of better building stone, the red ironstone of the Middle Lias, and the limestone of the Lower Oolite. The material employed in many of the churches on the other side of the county is much inferior. At Melton Mow-

Queniborough

bray the two stones mentioned above are blended together with a most pleasing effect. This church, indeed, is remarkably fine in many ways. It exhibits externally illustrations of all the Gothic styles. The lower stage of the tower is Early English, with the characteristic two-light windows ; the windows of the aisles range from Early English to late Decorated ; the large west window is Decorated, of the best period ; and the upper storey

Melton Mowbray

of the tower and the clerestory of forty-eight windows above the nave and transepts are Perpendicular. Nearly all of these may be seen in our illustration.

The churches at Thornton and Peatling Magna are especially noteworthy for the interest of their interiors, as is the fine church at Gaddesby, once belonging to the Knights Templars. In the large church at Lutterworth is a well-preserved fifteenth-century painting of the Doom, as well as numerous relics wrongly associated with Wycliffe, but nevertheless of much interest.

Leicestershire had a small number of ecclesiastical houses of different kinds, now mostly in a ruinous condition. Of the once important Augustinian Abbey

Ruins of Ulverscroft Priory

of Leicester little more remains than a few fragments of the walls, but another Augustinian foundation, Ulverscroft Priory, situated in the heart of the Charnwood Forest, has been better preserved. The remains of Grace Dieu nunnery are somewhat puzzling, being intermingled with the ruins of a Tudor house built on the same site. The Cistercian Abbey at Garendon has practically disappeared.

17. Architecture—(*b*) Military and Domestic

In early times all dwelling-places of any importance which were not built for purposes of religious worship had to be more or less strongly fortified: indeed some of the early churches themselves were obviously designed to serve, in emergencies, as places of refuge from hostile marauders. The non-ecclesiastical buildings, whether designed primarily for domestic or for military purposes, have naturally not survived the violence of man to nearly the same extent as have the churches themselves. Thus, of the once important fortified castles at Hinckley, Earl Shilton, Groby, and Mountsorrell, already referred to under the history of our county, practically nothing remains save the sites. Of Leicester Castle, the mound, which was probably erected at about the time of Alfred the Great, still exists. A plain brick building, some 150 years old, now stands on the site of the Great Hall, but fragments of late Norman work may still be seen in it. The castle and its neighbourhood, however, are now of greater interest to the antiquary than to the lover of architecture.

The ruins of the once fine castellated mansion at Ashby-de-la-Zouch, and the smaller mansion at Kirby Muxloe are more extensive. They are the remains of strongly-built and fortified houses, and are the only buildings in the county with any real claim for notice from the point of view of military architecture.

Ashby Castle was built in about 1474 by William, Lord Hastings (the Hastings of "Richard II"). The castle is familiar to most readers as the scene of the tournament in *Ivanhoe*, and it has a further interest from the fact that Mary Queen of Scots was imprisoned here in 1569.

Ruins of Castle at Ashby-de-la-Zouch

It was well built of sandstone, and was at one time a magnificent structure with stately towers. Garrisoned and defended for Charles during the great Civil War, it was at last evacuated and dismantled, and its ruins are now quite the most picturesque in the county. Kirby Castle also belonged to the family of Hastings, and was built at about the same date. The ruins of

Bradgate Hall, illustrated on page 140, are also of considerable interest, but little really remains save the shells of two towers, and traces of the walls, moat, and pleasaunces.

Of more recent date, and in a far better state of preservation, there exist within the county numerous fine

Belvoir Castle

examples of early domestic architecture. Foremost among these must be mentioned Belvoir Castle, for many centuries the home of the earls and dukes of Rutland. This finely situated building occupies a site on which has stood a castle since the Norman Conquest, and the site itself, an abrupt elevation, resembling an immense artificial mound, was adopted as a fortified stronghold even by the early Britons. Of the original or Norman building fragmentary remains only can now be seen in the base of the great Staunton Tower. The castle was allowed to go to ruin in the fifteenth century after the Wars of the Roses, and, having in the intervening years been rebuilt, was demolished by the Parliament in 1649 after the great Civil War. It was again rebuilt after the Restoration, and was again destroyed, this time by fire, in 1816, so that the present castle has not stood for much more than 100 years. It is a stately quadrangular building, with massive towers and walls of local ironstone and of white Lincolnshire limestone; it commands magnificent views up and down the Devon valley, and in its galleries hang many fine portraits and pictures by famous artists.

Other notable examples of domestic architecture are to be found in the numerous seats of the noble English families, who have made their homes in the county at different periods, particularly Quenby Hall, Staunton Harold, Donnington Hall, and Rothley Temple. Of these, Quenby Hall, built of brick on high ground, some 7 miles due east of Leicester, probably affords the finest example of Elizabethan architecture in the county. Another, and perhaps equally fine Elizabethan house, that at Carlton Curlieu, is of white stone, and other good examples of Tudor domestic architecture may be seen at Skeffington, about 10 miles east of Leicester, and at Brookesby, between Leicester and Melton Mowbray. Skeffington Hall was considerably enlarged in Jacobean times, but the original style was adhered to. There are several other examples of manor houses and halls of this period, and at Launde Abbey, in a remote part of East Leicestershire, some much

Quenby Hall

earlier work was incorporated with the handsome Tudor dwelling-house.

Of earlier date than the houses just considered may be mentioned an interesting fortified manor-house, surrounded by a moat, at Appleby Magna, in the extreme west of the county; this dates from the early fifteenth century, and there is a very fine example of a small house of still earlier date—thirteenth century—at Donnington-le-Heath near to Coalville. At Neville Holt, also, near to the Northamptonshire border, is an exceedingly picturesque and interesting house of various dates, from the fifteenth to the eighteenth century, and the monuments of some members of the great Neville family may be seen in the church which adjoins the house.

Lastly, there have survived in Leicester itself several specimens of medieval architecture. There is the old Town Hall, built in the fifteenth century, where it is very probable that Shakespeare himself played: there are the old Chantry House of William Wyggeston, and the much-altered Trinity Hospital in the Newarke, as well as two old gateways known as the Turret Arch or Rupert Gateway, and the Newarke Gate in the same quarter of the town. The picturesque Huntingdon Tower, which used to stand in High Street and was the last remnant of the great house of the Earls of Huntingdon, was removed for street improvements a few years ago.

The earliest dwelling-houses of any importance consisted of one large room—the Hall—on the ground floor (or somewhat above ground level as at Donnington), and suites of smaller rooms adjoining, the whole being strongly built and often defended by a moat. As time went on the need for defence grew less, and comfort was more aimed at, but for long the hall remained the

Jacobean Fireplace in the Old Town Hall, Leicester

principal room. Towards the end of the fifteenth century, and especially in the sixteenth, builders began

Street in an old quarter of Leicester
(*St Martin's Church in distance*)

to pay more attention to architectural effect, and the beautiful Elizabethan houses, several of which have been mentioned, were chiefly erected during this period. In these the hall is still a most important room ; the windows are generally large and square-headed, arches have become semicircular instead of pointed, there is often a tower over the main entrance, and the whole building is more elaborately ornamented. In the seventeenth century, however, everything became changed, and a favourite idea of the architect seems to have been to copy the Italian model. The finest example in Leicestershire of the work of this period is probably that in the north-east front of Staunton Harold, the seat of Earl Ferrers, situated near Ashby-de-la-Zouch. This is ascribed—probably correctly—to the great Inigo Jones, and is constructed of brick in the Palladian style, but the remainder of the house was not built until about the middle of the eighteenth century.

At Desford, near Leicester, there are two good brick-built houses, also of the seventeenth century, but of more homely design. Well-built houses of later dates are of course much more numerous. To the Gothic revival of the eighteenth century we owe the picturesque Donnington Hall, near Castle Donnington, formerly a seat of the Hastings family. Gopsall Hall is a good example of a more typical Georgian mansion.

The farms and cottages of the Leicestershire villages are often very pleasing, but do not exhibit many specially local characteristics. The old cottages in the Soar valley are often built of mud, and the local slate has been extensively used in the Charnwood district. Clay, and consequently bricks, are common in most parts, but many of the larger houses and village inns in the Tilton district are of the local warm marlstone. Timbered houses are neither so common nor so fine as in the adjoining county of Warwickshire, but at

Typical Leicestershire Cottages at Scraptoft

Ragdale, in the Wold district, the old hall was originally a half-timbered house, and together with the adjoining church makes a delightful picture.

18. Communications—(*a*) Roads

Direct historical evidence of the existence of roads in this country before the coming of the Romans is furnished by a sentence in one of Caesar's books, referring to " the well-known roads and paths " of the natives. But such evidence is scarcely necessary, for obviously the Celtic

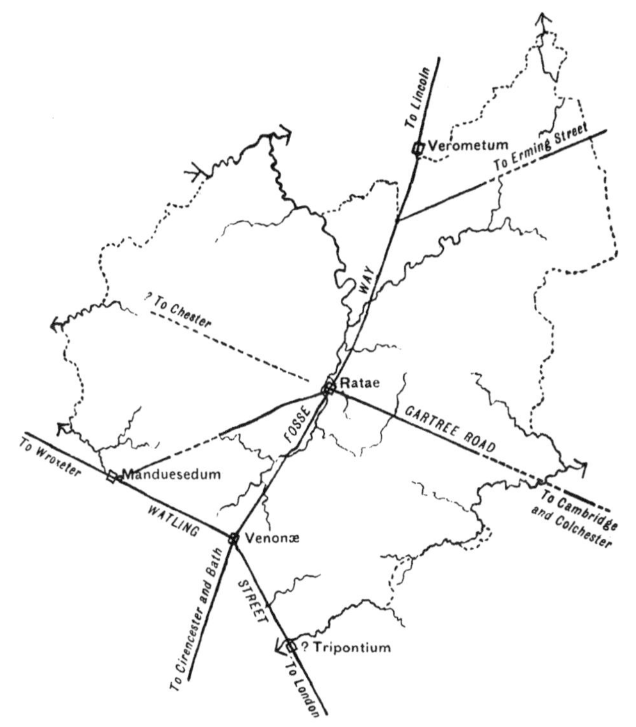

Roman Roads connecting with Leicester (Ratae)

inhabitants would have definite paths along which to travel when trading with the Continent, and with their kinsmen in distant parts of their own land. Moreover, since we know that they possessed well-made chariots, such roads in many places were probably carefully constructed. Some of these can be traced in the southern parts of England, and some of the most important are what have been called "ridgeway roads"—that is, they ran along the ridge between two river-basins and thus avoided the thick forests, marshy places, and river crossings. A good example of such a road in Leicestershire is that which runs from Six Hills to the north-east in the direction of Grantham.

On the coming of the Romans some of these old roads were taken as a basis for the construction of those wonderful highways which have persisted, often in good condition, to the present day. Both Watling Street and the Fosse Way, the Roman roads with which we are best acquainted in this part of the country, followed the line of older roads, straightening them out and overcoming many difficulties which the older tracks avoided. But although the Fosse Way, in a length of 182 miles, never departs more than 6 from the straight line, it should not be imagined that Roman roads are invariably quite so direct over long distances. Watling Street, for instance, the great military highway from London to the Welsh border, often deviates considerably from the straight line, as may be seen by a glance at our map of the Leicestershire roads.

The Fosse Way, so called from the ditches running beside it, passes through Leicester (Ratae) on its way from Lincoln to the south-west. In places it is now out of use, but it may still easily be traced for the whole distance through Leicestershire; indeed, in most parts

Disused portion of the Fosse Way near High Cross (Venonae)

The Fosse Way, nearer Leicester

it is an important modern highway. The present-day roads, however, are compelled to leave it after a greater or less interval, owing to the fact that nearly all of the important towns lie a few miles east or west of the old road. The road avoids high ground in this case by following the river valley until it leaves the county in the neighbourhood of Six Hills, whence ran a Romanised form of the old ridgeway road mentioned above.

Another less known Roman road, the Gartree Road, which afterwards came to be called the Via Devana, crossed the Fosse Way almost exactly at right angles at Leicester. It can still be traced almost to the border of the county, and again through a part of Northamptonshire and Huntingdon towards Cambridge and Colchester, all of which places are in an almost perfectly straight line with Leicester.

To the west or north-west the Gartree Road cannot be traced as a Roman road, though doubtless there was a track in early days which eventually gave rise to the modern highway to Ashby-de-la-Zouch and Burton-on-Trent. Most travellers to the north-west, however, probably preferred to deviate slightly to the left along the other road shown, and so meet the busy Watling Street near to Atherstone (Manduesedum), thus avoiding the wild and hilly district of the Charnwood Forest.

The construction of the more important Roman roads was most admirable ; in places they were raised six feet or more on embankments, while in others they were drained by parallel ditches, which, where they remain in Leicestershire, are nearly eight yards apart. Near Six

Roman Milestone which stood on the Fosse Way, near Leicester. (*Vide* p. 98)

Hills the surface, according to the eighteenth-century writer Stukeley, was constructed of red flints with their smooth sides uppermost, carefully set in a bed of gravel. Blocks of stone were used in other places in order to make a permanent track for the wheels of chariots, and some of the old stones are to-day used in Leicester as kerb stones; they may be seen, for example, in Granby Street.

The destruction of the Roman roads began, of course, very long ago, but chiefly took place during the great period of road-making in the late eighteenth and early nineteenth centuries, *i.e.* in the early coaching days. At this period most of the existing high-roads of Leicestershire, about 330 miles in all, were constructed. As has already been remarked, they constitute an excellent system of communication radiating from the central and most important town, and secondary roads, themselves quite well made, make a rough circuit of the county. The lanes and by-ways complete a close network over the whole county, and exhibit all gradations between well-constructed and tarred-surface roads in the populous districts, and green field-tracks or bridle roads connecting the more remote places.

19. Communications—(*b*) Canals and Railways

The conveyance of heavy merchandise by water rather than by road offers the important advantage of economy, but in olden days only a few favoured localities could benefit in this way. The idea of constructing locks, so as to conserve the water in otherwise non-navigable streams, seems to have originated about the year 1700, and in the following century many important canals were constructed in Leicestershire. One of the oldest, opened in 1779, was that by which the river Soar was made navigable from the Trent to Loughborough. It was extended to Leicester in 1791. The Charnwood Forest Canal was designed to carry loaded coal waggons from the Leicestershire coal district to a point near to Loughborough; the coal waggons were then hauled along an "edge-railway"—the first constructed in the world—to Loughborough, whence they continued their journey to Leicester by water.[1] The Charnwood Forest Canal has long been empty and disused, but numerous other canals were constructed during the early years of the last century; one of the most important being what is now called the Grand Junction Canal, which affords through communication with London and the Thames. This runs southward from Leicester, following the river Soar for the first few miles, and has a sharp rise of level—75 feet—at Foxton, near Market Harborough. The change of level was originally made by a flight of ten locks in rapid succession, but in 1900 these were replaced by an inclined railway, which raises or lowers huge tanks of water, each containing a barge and its contents. In these days, it is difficult to realise what an important part canals played before the coming

[1] The success attending some of these ventures may be gathered from the fact that original £100 canal shares at one time had a market value of £4800.

of railways. An old Leicestershire gazetteer of date 1846, quaintly refers to the "Fly-boats every night to London, in 54 hours, leaving goods at all the intermediate places," while daily boats also went from Leicester to places as far away as Liverpool, Leeds, Bradford, and

Canal Lifts on the Grand Junction Canal at Foxton

Hull, and three times a week to Exeter and all parts of the west. At the present time the canals of Leicestershire are by no means idle. Goods such as flour, as well as large quantities of coal, are regularly brought into Leicester by the canal route from the north.

The first true railways to be laid down in this country were the ancient "edge-railways," mentioned above. One of these was constructed in 1793 to carry coal to

Belvoir Castle from the canal a few miles away. It still remains, and is the oldest known example of a railway constructed for flanged wheels. The other system in these early days was the Outram-way, which in its oldest form consisted of iron plates, provided with an

Bronze Ticket used on the Leicester and Swannington Railway

edge to keep the wheels of the waggon on the track.[1] A portion of the old Outram-way, laid in 1799, still exists near Ashby-de-la-Zouch.

On the introduction of steam-power for railways Leicestershire again was among the earliest pioneers.

[1] The plate-layers of our modern railways are so called because their predecessors really did lay plates for the track.

The old Leicester and Swannington Railway, now long since absorbed into the Midland Railway, was con-

Swannington Incline, on the old Leicester and Swannington Railway

structed by George Stephenson and his son Robert in 1830–1834. It was designed to bring coal direct to Leicester, and succeeded in breaking down the monopoly of the canal companies. The commercial prosperity of Leicester may be said to date from its completion, and the consequent abundance of cheap coal which it brought into the town.

At the present day Leicestershire is well provided for by the lines of a number of different railway companies, most of which, like the main roads, converge towards the town of Leicester. The main Midland route both to Manchester and to Scotland passes through Leicester, and an important branch of this railway also crosses the country from west to east, providing a direct route from Birmingham to the east coast. Another branch of the Midland passes through Coalville and Ashby-de-la-Zouch to Burton-on-Trent, and yet another joins the L. and N.W. Railway main line at Rugby.

The Great Central main line from Sheffield to London, completed in 1900, also passes through the county from north to south, and has important stations at Loughborough and Leicester, while branches of both the Great Northern and L. and N.W. Railways run into the latter town from their main lines, which lie respectively to the east and west of the county. The L. and N.W. Railway has also a branch from its main line station at Nuneaton which passes through Market Bosworth to the Leicestershire coalfield; much coal proceeding by this route direct to London. East Leicestershire is not quite so well served, but a joint line belonging to the G.N. and L. and N.W. Railways passes north and south through Melton Mowbray.

20. Administration and Divisions

It is in early Anglo-Saxon times that we find the real beginnings of our present-day system of government. The Teutonic settlers arranged themselves into townships or "vills," the public affairs of which were discussed by meetings of the freemen of the community. The townships were grouped into larger divisions, known as "hundreds," perhaps so called from their being composed of the lands of one hundred families. At the head of the Hundred was the Hundred man, and by him the "moot," a monthly meeting of the principal men from each township, was summoned. The duties of the Hundred Moot, which thus corresponded to our modern District Council, were to levy taxes, to try criminals, to exact fines, and to attend to matters of military organisation in its district. As time went on they were gradually shorn of these powers and became of less importance, until now we do not even know what were the boundaries of the original Leicestershire Hundreds. This, however, is partly due to the fact that, at some time during the Danish period, other divisions were instituted, known as Wapentakes: these were much larger than Hundreds, and probably more closely related to military organisation. They are only found in those parts of England in which the Danes established themselves, and in our own county seem to have become confused at some later period with the real Hundreds. At the Domesday Survey Leicestershire is described as consisting of four

Wapentakes—Guthlaxton, Goscote, Gartree, and Framland, the first three of which met at Leicester. In later years two of these, Guthlaxton and Goscote, were subdivided, the former into Guthlaxton and Sparkenhoe, and the latter into East and West Goscote. There is in the Domesday Book no mention of the smaller Hundreds, but that there were such divisions we know from another valuable document, "The Leicestershire Survey," which was compiled between 1124 and 1129, during the reign of Henry I. In this we find the first clear reference to different Hundreds, but these small divisions in no way correspond to the Wapentakes.

The Hundred Moots were usually held at some important hill or rock or tree. These survived for many years in the ancient Swanimote Courts of the Charnwood Forest district, which were held at Copt Oak, at the Swanimote Rock near Whitwick (where there was also a Swanimote Oak), and at a place not far from Ive's Head.

In addition to the Hundred Moots, there came into existence, with the creation of the shires or counties, the Shire Moot, one of the duties of which was to settle disputes between people in different Hundreds. The chief officers in the Shire Moot were the Shire-reeve or Sheriff and the Ealdorman or Earl: these were appointed by the King or the National Council (Witan), but the other members were chosen by the Hundred Moots. In Leicester itself it seems that at the close of the Saxon period the burgesses enjoyed a considerable degree of freedom in their local government. They paid a yearly

The Swanimote Oak, near Whitwick

tax to the King in coin and honey, and twelve of them had to accompany him on military expeditions.

At the coming of the Normans the township automatically became the manor, but there was really no very great change in the method of organisation of the affairs of the district. In 1360 the first Commissioners or Justices of the Peace were appointed, and from that date until comparatively recent times the Justices acted both as magistrates and administrators of public affairs.

In 1888 a great change came about, closely following on the grant of the parliamentary franchise to agricultural labourers a few years earlier. The old system was changed, and the country was divided into sixty-two administrative counties, each with an elected County Council which at once acquired control of most of the affairs of the county. This change was followed a few years later by another, which resulted in the establishment of elected District (both Urban and Rural) and Parish Councils.

The boundaries of the administrative county of Leicestershire, with one or two slight deviations, follow those of the ancient county. It is divided into fifty-four electoral divisions, each of which returns one councillor, and these, with the eighteen county aldermen, constitute the County Council. The County Council is responsible for the local government of the whole of the county with the exception of the City of Leicester.[1] The Justices, of course, still retain authority in judicial affairs

[1] Leicester was made a City after the visit of the King and Queen in 1919.

and hold their Courts of Quarter Sessions at Leicester, as well as Petty Sessions at the lesser towns of the county. Only the most serious offences, such as murder and treason are outside their jurisdiction, and these go before the Judge of the Assize.

For Poor Law legislation, which dates from the time of Queen Elizabeth, the county is divided into eleven Poor Law Unions, each of which is under the control of a Board of Guardians. In urban districts these are specially elected, but in rural districts a district councillor is also the representative of his electoral area on the Board of Guardians. Lastly, the unit in the administrative affairs of the county is the Civil Parish, of which there are some 324 in Leicestershire. The rural Parish Councils, besides supplying members to the District Councils, have certain important powers of their own, but some of these powers are seldom exerted.

The Rural District Councils, among other duties, have charge of the sanitary administration of their district, and are also responsible for the proper maintenance of all roads which are not main roads. The Urban District Councils have similar powers, and for large districts are practically equivalent to, though they have not the dignity of, the Town Councils of Municipal or County Boroughs. In Leicestershire there are twelve rural districts and eleven urban districts. Loughborough is a Municipal Borough, *i.e.* has received a charter of incorporation from the Crown, and possesses a mayor, six aldermen, and eighteen councillors, but it also supplies members to the County Council. Leicester itself, on the other hand, being a County Borough, is exempted from the authority of the County Council, and its civil administration is for the most part in the hands of its City Council, consisting of mayor, sixteen aldermen, and forty-eight councillors. These work

The Old Grammar School, Market Harborough

through sub-committees appointed to deal with separate matters, such as education or lighting. The old grammar schools of the Market towns, as well as the Wyggeston grammar school at Leicester, were formerly independently managed, but now practically all are under either the Leicester or County Education Authority.

We thus see that for practically all purposes save the administration of justice, local affairs are in the hands of the people, whether dwellers in the town or country: indeed the measure of local government enjoyed by the people in this country is much greater than in most other European States, though in such matters as education, the central departments wisely retain the power of withholding the Exchequer grants, if the work is not done to their satisfaction.

For Parliamentary elections Leicestershire is divided into four districts, each of which returns one member, while Leicester itself is now (1919) represented by three members. Lastly, for purposes of ecclesiastical administration, the county forms an Archdeaconry, consisting of seven deaneries, in the diocese of Peterborough. Formerly (*i.e.* until 1839) Leicestershire was attached to the See of Lincoln.

21. Roll of Honour

In the early days of English history the small town of Leicester held, as we have seen, a relatively very important position in the affairs of the kingdom, so that it is not remarkable to find the names of distinguished persons connected with it. Simon de Montfort, although born in France, came early to this country to enter into the estates belonging to the Earldom of Leicester, which he inherited from his grandmother. His part in the history of our town and county has been already touched upon, but it is of further interest here to note that he is one of the four benefactors of the town who are commemorated by the statues at the four corners of the Clock Tower. John of Gaunt, too, who also was an Earl of

Lady Jane Grey

Leicester, has been mentioned; he probably spent more time in the town than did de Montfort.

Another royal personage, Lady Jane Grey (1537–1554) was a native of the county, and in some respects her name is the most interesting on our list. The ruins of Bradgate

Hall, her birthplace, are illustrated below, and it was here that Roger Ascham found her, as a child of thirteen, reading Plato under the trees.

Wolsey, the great Cardinal of Henry VIII, was associated with Leicester only in his last days. On his

Ruins of Bradgate Hall

way from the north to attend his trial, he was taken ill, and died at Leicester Abbey. He was buried in the chapel of the Abbey, but at its dissolution the whole building was demolished, and antiquaries have searched in vain for any trace of his tomb.

At Brookesby Hall, a fine old building still standing close by the main road from Leicester to Melton Mowbray, was born yet another who played an important part in the affairs of State of his time, George Villiers, first Duke of Buckingham (1592–1623); and at Noseley, in the south-east of the county, lived the great parliamentarian and soldier, Sir Arthur Hesilrige or Haselrig, a friend of Cromwell, and one of the five members who were imprisoned by Charles I after the first important breach between that monarch and his Parliament.

Turning to dignitaries of the Church we find a long list of names, some certainly of the first magnitude, in the county's roll of honour. Foremost among these in point of date comes Grosseteste, Archdeacon of Leicester, and afterwards Bishop of Lincoln (1175–1253), who was, like de Montfort, an opponent of Henry III. First in importance, however, must be mentioned the great reformer, John Wycliffe (*c.* 1325–1384), who became rector of Lutterworth in 1374. Simple, unostentatious, and a fearless preacher, he gained great popularity. While at Lutterworth, he dared openly to deny the doctrine of Transubstantiation, and so powerful were his supporters that he survived the ensuing attacks made upon him. It was from here, too, that he sent out his band of "poor preachers," equipped with his own English translation of the Bible.

Another great churchman, the martyr—Bishop Hugh Latimer—was born at Thurcaston about 1490, and was in more ways than one a follower of Wycliffe. He first came to notice by his refusal to deny the doctrines of Luther, and for long encouraged Puritanism in his diocese (Worcester). He was finally condemned as a heretic, and burnt at Oxford in 1555 with Cranmer and Ridley.

Although not a native of the county, Archbishop Laud

Hugh Latimer, Bishop of Worcester

(1573–1645) was connected with it as being for some years rector of Ibstock, and among the many well-known Nonconformist divines who have belonged by birth to Leicestershire must be named George Fox, the founder of the Society of Friends, born at Fenny Drayton in 1624; and the great Baptist preacher, Robert Hall, who was born at Arnesby in 1764, and spent many years at Leicester.

Leicestershire can boast of no distinguished artists, and of but few great poets or dramatists, though the greatest writer of lyrics, Robert Herrick (?1591–1674) came of a Leicester family, as did also Francis Beaumont, one of the greatest of Elizabethan dramatists, who was born at Gracedieu in 1584, and died in 1616. George Crabbe, the poet (1754–1832), was some time curate of Stathern, and afterwards rector of Muston, at the end of the eighteenth century. But there is no lack of other writers. Robert Burton, "whose company was very merry, facete, and juvenile," was born at Lindley Hall in 1577, and is best known as the author of that wonderfully learned farrago, the "Anatomy of Melancholy." He was for ten years rector of Seagrave. William Burton, his brother, is not so well known as a writer, but as an antiquary and historian of his native county he is of especial interest to us. His "Description of Leicestershire," published in 1622, is a monumental work, and was one of the very earliest of the English county histories. It is said to have inspired Dugdale's "Warwickshire," which was published some thirty years later.

Dr Johnson, though not a native of the county, taught for a short time at Market Bosworth grammar school, and the same school was for several years the scene of the activities of the great scholar, Anthony Blackwall. Mrs Barbauld, in her day a famous writer of books for children, was born at Kibworth Beauchamp in 1743; and Lord Macaulay (1800–1859) who was born at Rothley, was renowned as a historian, essayist, politician, talker, man of letters, and verse writer. He was created Baron

Lord Macaulay

Macaulay of Rothley in 1857, but he does not appear to have had any very close connection with his native county during his later life.

John Manners, Marquis of Granby, son of the Duke of Rutland, was a distinguished soldier, and served with great credit in Flanders in the middle of the eighteenth century as Colonel of the Leicester Blues, becoming Commander-in-Chief of the British Army in the Seven

John Manners, Marquis of Granby

Years' War. Another well-known soldier and cavalry leader was Colonel Burnaby, who is best known by his "Ride to Khiva" (1875). He was killed in 1885 in the attempt to relieve Khartoum.

In the realm of natural science and philosophy the roll of honour is not a long one, but two great naturalist travellers were connected with the county. These were Alfred Russel Wallace, the co-discoverer with Darwin of the theory of evolution, who taught at a Leicester school, and his distinguished friend, H. W. Bates, the traveller of the Amazons (*b.* Leicester 1825, *d.* 1892) who discovered over 8000 species of animals and plants. In earlier days also, Thomas Simpson (1710-1761) was a distinguished mathematician, who hailed from Market Bosworth, and to still earlier times (seventeenth century) belonged Lilly, the astrologer, a native of Diseworth (1602–31), who, though not a man of science, was a great authority on astrological literature.

Lastly must be mentioned Thomas Cook, Baptist and temperance advocate, who in 1841 organised the first publicly advertised excursion by train, which ran from Leicester to Loughborough and back. From this humble beginning arose the great tourist and excursion business now well known throughout the world.

22. The Chief Towns and Villages of Leicestershire

(The smaller villages are only mentioned when possessing features of more than ordinary interest.)

(The population in 1911 is given in brackets after each name.)

Anstey (2976) is a large and pleasantly situated industrial (boot and shoe-making) village about 4 m. N.W. of Leicester, with which it is connected by a broad lane said to be of Roman origin. (pp. 21, 70.)

Appleby Magna and Parva (675) are in the extreme west of the county, 5 m. S.W. of Ashby-de-la-Zouch. At the former is a fine Dec. church and an old moated house. At Appleby Parva is an old grammar school, designed by Sir Christopher Wren. (p. 116.)

Arnesby (354), 8 m. S. of Leicester. St Peter's Church has a good deal of Norman work still remaining. (pp. 104, 143.)

Ashby-de-la-Zouch (4927), on the western border, is a pleasant little town in spite of its situation in the Leicestershire coalfield. It came into the possession of the Zouch family shortly after the Conquest, but since the middle of the fifteenth century has belonged to the Hastings family. Its castle was the scene of the Ivanhoe tournament, and at a later date Mary Queen of Scots was imprisoned there: in the Civil Wars it remained for long an important Royalist stronghold. The parish church (St Helen's), which is close to the castle, is a good example of the Perp. style, and contains some rare furniture and very fine monuments. (pp. 18, 33, 87-8, 107, 111-2, 124, 129, 131.)

Aylestone, though now a suburb of Leicester, retains much of its old picturesque village aspect, including the old Tudor Hall, at which Charles I lodged during the siege of Leicester. Its large church (St Andrew's) in the E.E. and Dec. styles, has a chancel which is larger than the body of the church, as well as a very unusual broach spire.

Barkby (670) and **Beeby** (95), small neighbouring villages, which are prettily situated about 4 m. N.E. of Leicester: Barkby has a very interesting church (St Mary), mostly in the E.E. and early Dec. styles, with a fine broach spire. All Saints', Beeby, has a peculiar unfinished spire, which has given rise to various legends. (p. 105.)

Barrow-on-Soar (2481), an industrial village on the right bank of the Soar, 3 m. S.E. of Loughborough. Many interesting fossils from the Lower Lias have been obtained in the neighbourhood. Here are the kennels of the Quorn Hunt. (pp. 31, 33, 77-8, 98.)

Barwell (2998), a large and uninteresting manufacturing village 2 m. N.N.E. of Hinckley, with a rather fine church in the E.E. and Dec. style.

Billesdon (594), about 8 m. E. of Leicester, has long been a famous hunting centre. At its old school two Leicestershire celebrities, George Villiers, Duke of Buckingham, and George Fox are said to have been educated. A prominent eminence in the neighbourhood is Billesdon Coplow, which rises to a height of 700 feet, and is easily visible from many points on the other side of the county. (pp. 21, 44, 94.)

Birstall (751) is a secluded little village, 3 m. N. of Leicester. The church of St James has a Saxon window. (p. 103.)

Blaby (1959), an industrial village and centre of a Poor Law Union and Rural District, 4 m. S.S.W. of Leicester.

Bottesford (1174), in the extreme N. of the county, in the plain of the River Devon, possesses a handsome village church (St Mary), chiefly in the Perp. style, with the highest spire in the county (212 feet). There are in the church some fine monuments to the families of de Ros and Manners, and the earls of Rutland, whose home, Belvoir Castle, is 3 m. to the S. Remains of the stocks stand in the village, and an ancient market cross. (p. 11.)

Breedon-on-the-Hill (752), a quarrying village on the Derbyshire border of the county. The rock here is of the Mountain Limestone formation, which is met with scarcely anywhere else in the county. The abrupt and conspicuous hill was the site of a very ancient earthwork of considerable extent, and the old priory church contains some most interesting work, including a Norman font and a sculptured frieze, with representations of birds and animals, which is believed to be of Saxon date. (pp. 52, 77, 103.)

Billesdon Village and Church

Burrough-on-the-Hill[1] (200) is on the marlstone escarpment of East Leicestershire, 6 m. S. of Melton Mowbray. On the summit of the hill are ancient earthworks, probably of British origin: the view from this point is very extensive. (pp. 16, 32.)

Castle Donnington (2529) is a pleasant little town, 9 m. N.E. of Ashby-de-la-Zouch, at the N.W. apex of the county, overlooking the Trent. No vestiges of the castle remain, but the church of St Luke shows beautiful work in both early and late Dec. styles. At one time, a quite busy inland port was situated here, goods which had been brought up the Trent being landed to proceed southward by road. Donnington Park, close by, also has a fine situation overlooking the river, and was until recently a seat of the Hastings family. It has some very fine oaks, notably "Chaucer's Oak," which is 44 feet in girth. (pp. 12, 94, 104.)

Coalville (8756) is a colliery town, 6 m. E.S.E. of Ashby-de-la-Zouch, on the western edge of the Charnwood Forest district. Bricks and tiles are made here. (pp. 73-4, 131.)

Congerstone (195) is about 3 m. N.W. of Market Bosworth. The remains of an ancient gibbet, said to be the last remaining in England, may be seen by the side of the road from this village to Sibson. Gopsall Hall, close by, is a stately Georgian mansion where Handel was a visitor, and is said to have composed his "Messiah."

Croft (742) is a picturesque village, 8 m. S.S.W. of Leicester, on a branch of the Upper Soar which has here cut its way through the granite. Croft Hill, also of granite, is an isolated eminence from which extensive views may be obtained over the surrounding low-lying country. There are large steam granite quarries. (pp. 72, 76.)

Croxton Kerrial (452), 9 m. N.E. of Melton Mowbray, in the N.E. of the county, and close to the Lincolnshire

[1] Also spelt Burrow.

boundary, stands high, and commands fine views, one of which includes Belvoir Castle, some 3 m. away. The fine church of St John Baptist is chiefly in the Dec. style, and has good carved seats. Many of the houses are built of the local limestone. At Croxton Park there is an annual race-meeting, which usually terminates the fox-hunting season. (pp. 14, 32.)

Dalby-on-the-Wolds (368) or Old Dalby, is a beautifully situated village close by the Fosse Way, some 12 m. N. from Leicester. (p. 11.)

Donisthorpe (2444) is a colliery district, 3 m. S. of Ashby-de-la-Zouch, formerly in Derbyshire, and transferred by the Act of 1888.

Earl Shilton (4190) is a large industrial township, 9 m. S.W. of Leicester. The earls of Leicester once had a small castle here, and the Manor still belongs to the Duchy of Lancaster, with which the Earldom of Leicester first became united in 1265. (p. 111.)

Ellistown (2362) is a modern colliery village, 5 m. N. of Market Bosworth, which has extensive brick and fire-clay works.

Enderby (2667) is a large village 5 m. W. of Leicester, where are important granite quarries. (p. 73.)

Evington (958) is a picturesque village, which has now become almost a suburb of Leicester. Ancient earthworks may be easily traced in the fields close by the old parish church, which contains some rare fragments of window glass. (p. 105.)

Grimston (176) is a remote village on the Wolds, 5 m. W.N.W. of Melton, where the old stocks may still be seen on the village green. There is also an old cross in the churchyard.

Groby (910), 4½ m. N.W. of Leicester, has important syenite granite quarries, and a picturesque old manor-house. Of the old Norman castle only the mound now remains. Elizabeth Woodville, afterwards wife of Edward IV, lived here for many years. (pp. 23, 40, 111.)

Hallaton (566), some 12 m. S.S.E. of Leicester, is an old-world village which was once a market town. There is an ancient earthwork, and the fine church, dedicated to St Michael, is in E.E. and Dec. styles, and retains some of the older Norman arches. An old Easter custom, which still survives, takes the form of a scramble for a hare pie. (pp. 94, 104-5, 107.)

Hathern (1209) is a finely situated village, 3 m. N.W. of Loughborough, overlooking the Soar valley. There is a good village cross. Hosiery is still made on the hand-frame knitting machines.

Hinckley (12,837) is a small but busy manufacturing town some 13 m. S.W. of Leicester. Hosiery-making has been the chief occupation ever since the introduction of the stocking frame here by William Iliffe in 1640, but boots and shoes are also made. In medieval times the town was a borough, and in still earlier days there was a Norman castle which stood at the top of the present Castle Street. This is supposed to have been destroyed during the Wars of the Roses, but a portion of the moat still exists. The old church, St Mary's, which is one of the largest in the county, is a cruciform building, mainly Dec., with embattled western tower and lofty spire : it has been much restored. (pp. 8, 58, 66, 71-2, 80, 111.)

Hugglescote (5659) with Donnington, is a small mining town, 5 m. S.E. of Ashby-de-la-Zouch, almost on the slopes of Bardon Hill.

Ibstock (4946), a large mining village in the Leicestershire coalfield, 6 m. S.E. of Ashby-de-la-Zouch. (pp. 54, 142.)

Kegworth (2220) is a manufacturing village close to the Nottinghamshire border, 6 m. N.W. of Loughborough; it has a beautifully designed church (St Andrew) in the Dec. style. Thomas Moore, the poet, resided here.

Kibworth (Kibworth Beauchamp and Kibworth Harcourt) (1807) together form a large village on the main road between Leicester and Market Harborough, 6 m. N.W. of the latter place. (p. 144.)

Kirby Muxloe (1032) is a residential village about 4 m. W. of Leicester. The old castle, or rather castellated mansion, is still a fine specimen of fifteenth-century building, but the ruins have suffered from neglect during recent years. (p. 111.)

The Langtons are a group of five small villages, distinguished by the prefixes Church, East-, Tur-, Thorpe-, and West-, about 12 m. to the S.E. of Leicester.

Leicester (227,222) is the county town, and one of the largest manufacturing towns in the Midlands. The relics of the Roman occupation now remaining are somewhat scanty, but nevertheless of great interest. The old castle has been replaced by a brick building, but the Great Hall, together with other fragments of the Norman building, have been embodied in the present structures. Of the six parish churches mentioned in the Domesday Survey, four remain, or at least their sites are still occupied by ancient churches bearing the same names. These are St Martin's, St Margaret's, St Nicholas's, and All Saints'. St Martin's (chiefly E.E.) is the civic church, and is as irregular in plan as St Margaret's is symmetrical. All Saints' possesses an old clock with quarter jacks over the S. door, first erected in the reign of James I. St Mary's is perhaps the most interesting of them all. It was originally the church of St Mary de Castro, built in 1107, and among other peculiarities its plan gives the appearance of being that of a double church, and indeed at one time the building served a double function as a parochial and collegiate church. The modern portions of the town are chiefly brick-built, and give the impression of a well-kept and busy manufacturing town. Its centre is the Clock Tower, which stands just outside the East Gate of the old town (*cf.* plan on page 87).

There are some good modern buildings, among them the Town Hall and Parr's Bank, and the residential suburbs are well supplied by electric tramways radiating from the

The Castle and St Mary's Church, Leicester

Clock Tower. The principal school of the county, which owes its foundation to a sixteenth-century endowment, is named after the benefactor, William Wyggeston, and is situated in the old part of Leicester, adjoining both St

Martin's churchyard and the interesting old town hall. (pp. 3, 26, 57-8, 66, 68, 80-9, 95-8, 102-4, 116-8, 124, 126, 128, 133, 135, 138-40, 146.)

Loughborough (22,990), the second largest town in the county, is situated some 10 m. N. by W. of Leicester, and within easy reach of the Charnwood Forest; it is a busy

St Mary's, Lutterworth
(*John Wycliffe became Rector in* 1374)

manufacturing centre, the chief industries being hosiery-making and engineering. There is also a famous bell foundry, where the great bell in St Paul's Cathedral—Great Paul—was cast, and whence, being too large to go by rail, it was transferred by road to London. The most interesting building is the large parish church of All Saints', a cruciform structure in the Dec. and Perp. styles. (pp. 6, 11, 24, 58, 61, 70-1, 107, 127, 136.)

Lutterworth (1896), a small market town, some 13 m.

S. by W. of Leicester, is pleasantly situated on a slope rising from the little river Swift. The large church of St Mary the Virgin, with which John Wycliffe was for many years associated, is the chief point of interest. The E.E. tower was originally surmounted by a spire, but the latter fell during a great gale in 1703. A fine model of the church is among the very interesting collection of relics preserved in the edifice. (pp. 22, 107, 109-10, 141.)

Market Bosworth (729), situated 13 m. W. of Leicester, was once a more important town than it is at the present day. The church of St Peter has a lofty steeple, which forms an easily noticeable landmark for a wide area. In the old park, through which passes the main road from Leicester, is a clump of trees marking the spot at which Richard raised his standard before the fatal battle of Bosworth Field, in 1485. The battle itself, which terminated the Wars of the Roses, was fought at a point some 2 m. to the S., and the battlefield is best reached from Shenton Station. Dr Johnson was in 1732 a master at the grammar school here. (pp. 84, 131, 143, 146.)

Market Harborough (8853), on the Northamptonshire border of the county, was originally best known as a hunting centre, but is now becoming of more and more importance as a manufacturing town: its population includes that of Great and Little Bowden. Traces of Roman occupation have been discovered, and the fine church, dedicated to St Dionysius, is said to have been built by John of Gaunt; it has a beautiful Dec. steeple. Harborough also figured more than once in the Civil War. The old grammar school, no longer used as such, was built in 1618, and stands on wooden pillars, with a small market-place beneath. (pp. 68, 86, 105, 137.)

Measham (2303), a large colliery and brick-making village in the extreme W. of the county, 4 m. S.S.W. of Ashby-de-la-Zouch: like its neighbour, Donisthorpe, it was formerly in Derbyshire.

Melton Mowbray (9202), a pleasant old town and famous hunting centre, some 14 m. N.E. of Leicester. The church (St Mary's) is in many ways the finest in the county; it was probably commenced by Wm. de Melton, an Archbishop of York, in the early fourteenth century. Close by is an old house said to have been occupied by Anne of

A Meet of the Atherstone Hounds in the Market Place at Market Bosworth

Cleves, which previously belonged to Thomas Cromwell. Melton has for long been famous for its pork pies and Stilton cheeses. (pp. 45, 65, 86, 103, 105, 107, 131.)

Mountsorrel (2491), a very long village on the main road between Leicester and Loughborough, 4 m. S.E. of the latter. It was once the site of a Norman castle, and later a market town, but is now chiefly famous for its extensive granite quarries. (pp. 36, 76, 94, 111.)

Newbold Verdon (or **de Verdun**) (1064), a village about 10 m. W. of Leicester. The old Queen Anne Hall (now used as a farm-house), in which Lady Mary Wortley Montagu once resided, is still partly surrounded by a well-filled moat.

Oadby (2609), 3½ m. S.E. of Leicester. The Dec. church of St Peter is a fine building with a hexagonal broach spire. (p. 107.)

Quorndon (2363) or **Quorn**, a manufacturing village on the main road between Leicester and Loughborough, 2½ m. S.E. of the latter, is famous as the birthplace of the Quorn Hunt. (p. 43.)

Ratby (2112) is a rather uninteresting manufacturing village about 5 m. W.N.W. from Leicester. Close by are the interesting earthworks at Bury Camp and Ratby Burroughs. There is also an old moated farm-house and a spring called the Holy Well, the water from which is said never to freeze. (p. 4.)

Rothley (2006), a picturesque village on the edge of the Charnwood Forest, is rapidly becoming more important as a residential district. Rothley Temple, where Lord Macaulay was born, was founded by the Knights Templar in the reign of Henry III, but the present building is mainly of Elizabethan date. There is an ancient cross, probably Saxon, in the churchyard. (pp. 98, 103, 114, 144.)

Shepshed (or **Sheepshead**) (5542), 5 m. W. of Loughborough, a large and partly industrial parish which includes the ecclesiastical parish of Oaks in Charnwood, and a considerable portion of the Charnwood Forest.

Sileby (3082), a manufacturing village near to the right bank of the Soar, 8 m. N. of Leicester.

Stoke Golding (613) 5 m. S. by W. of Market Bosworth, and not far from Bosworth Field ; here indeed the battle terminated. There is a small but beautiful Dec. church, dedicated to St Margaret.

Stoughton (136), a small and pretty village about 3½ m. E.S.E of Leicester. St Mary's is a handsome Dec. church,

The Old Churchyard Cross, Stoughton

and there is a good churchyard cross. Stoughton Grange was once the principal grange of Leicester Abbey. (p. 105.)

Swannington (2050), a colliery village, which gave its name to the old Leicester and Swannington Railway, opened in 1833. (p. 130.)

Syston (3087) a large industrial village, 5 m. E. by N. of Leicester. (pp. 70, 94.)

Thurcaston (345) is a picturesque and thoroughly rural village about 4 m. N. by W. of Leicester. Bishop Hugh Latimer was born here in *c.* 1485. There are several old houses in the village, one of which is said, on somewhat doubtful authority, to be his birthplace. (p. 141.)

Tilton (130), a small and picturesque village, 10 m. E. of Leicester, on the East Leicestershire marlstone, about 700 feet above sea level. St Peter's is a fine old Norm. and Perp. church, and there is an ancient cross in the churchyard. (pp. 16, 32, 50, 94.)

Waltham-on-the-Wolds (543), a pretty village, 5 m. N.E. of Melton Mowbray, was once a place of more importance and a market town for the surrounding district. St Mary's is a good E.E. cruciform church, which retains fragments of late Norm. or Trans. period, and has what is perhaps the best font in the county. (p. 104.)

Whitwick (4133) is a mining village, 5 m. E. of Ashby-de-la-Zouch, close to the wildest part of the Forest district. It was once a market town and head of an important manor. Close by is the modern Cistercian Monastery of St Bernard, founded in 1835, designed by Pugin, where a large hoard of Roman coins was found in 1840. (pp. 17, 74, 133-4.)

Wigston Magna (8650), is an industrial village 4 m. S. of Leicester; it is often called "Wigston Two Steeples," from having two churches, St Wolstan's and All Saints', both of Dec. period. (p. 70.)

Woodhouse (1458) and **Woodhouse Eaves** (1199) are two pretty villages close to Quorn on the "Eaves" or edge of the Charnwood Forest. They are well-known holiday resorts for the people of Leicester and Loughborough. Beaumanor Park, where the Herrick family lived for many years, and Beacon Hill, the second highest point in the county, are close by.

Wymondham (626), near the source of the River Eye, 6 m. E. of Melton Mowbray, and once a more important place. Like Waltham, a few miles away, it has a fine E.E. church, dedicated to St Peter, with traces of Norman work. There is an ancient churchyard cross, and a picturesque old grammar school, founded in 1637. (p. 13.)

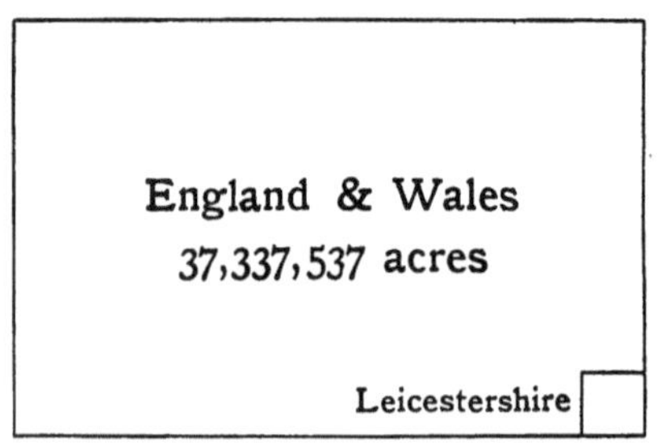

Fig. 1. Area of Leicestershire (530,642 acres) compared with that of England and Wales

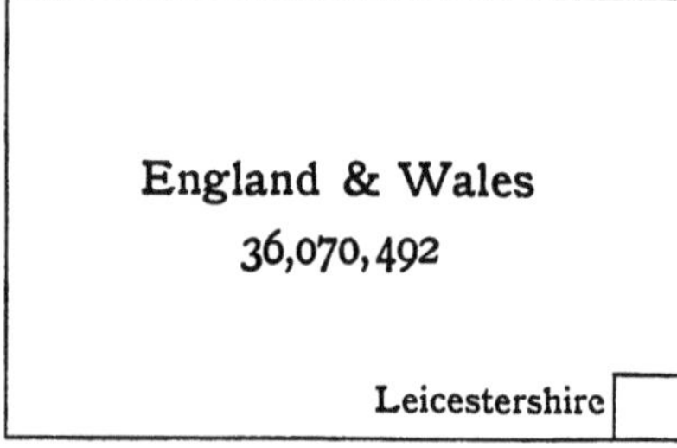

Fig. 2. Population of Leicestershire (476,553) compared with that of England and Wales in 1911

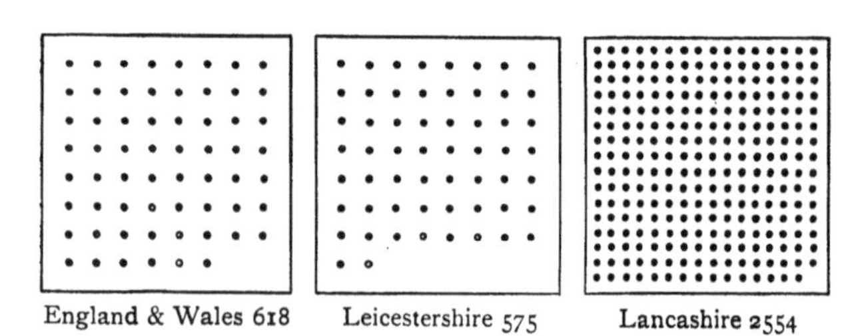

Fig. 3. Comparative density of Population to the square mile in 1911. (Each dot represents 10 persons)

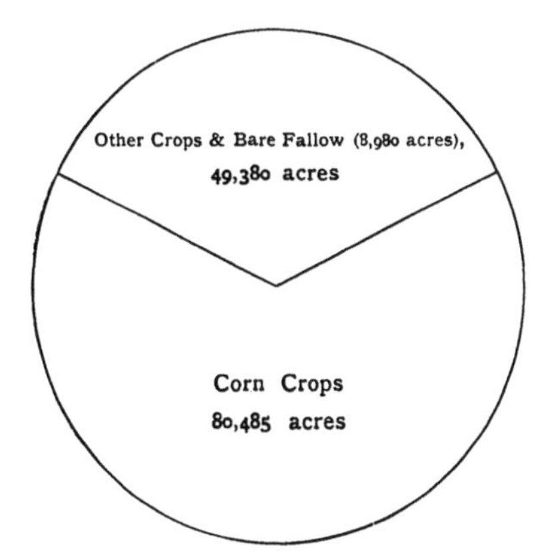

Fig. 4. Proportionate area under Corn Crops compared with that of other cultivated land in Leicestershire

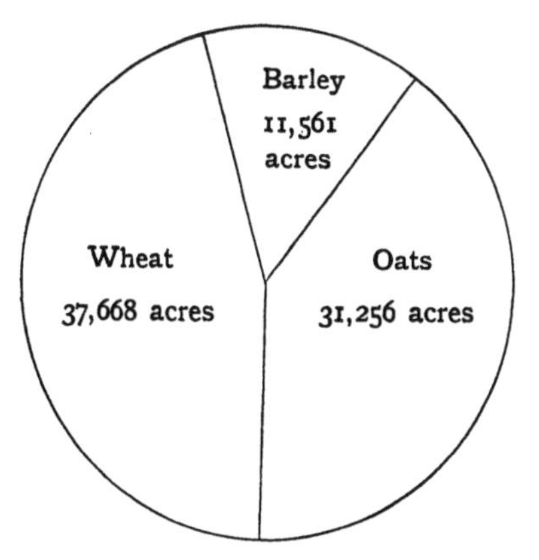

Fig. 5. Proportionate areas of chief Cereals in Leicestershire

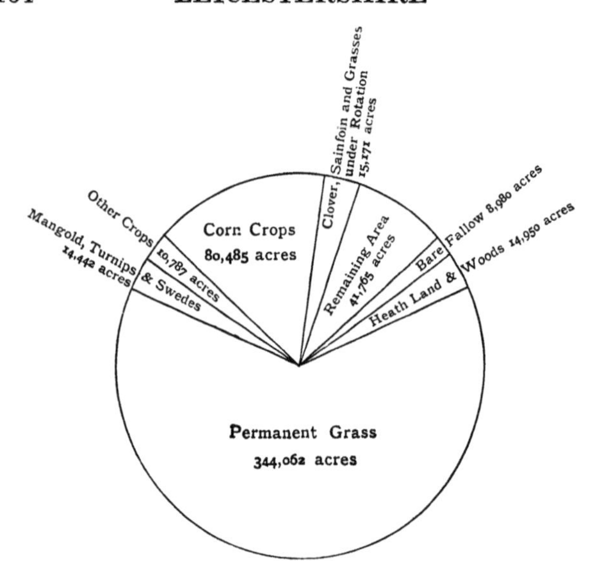

Fig. 6. Proportionate areas of Land in Leicestershire

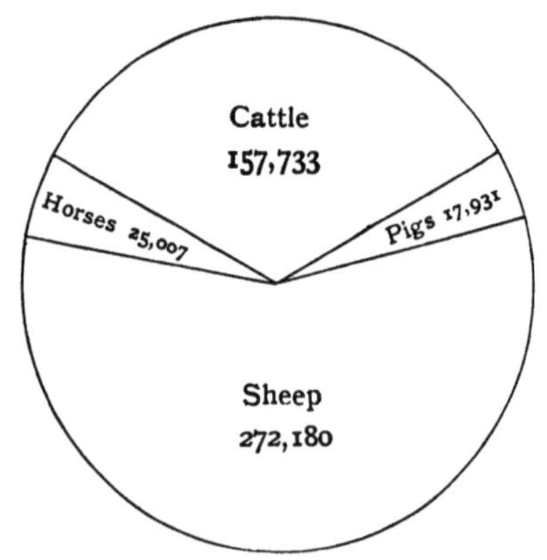

Fig. 7. Proportionate numbers of Live Stock in Leicestershire

www.ingramcontent.com/pod-product-compliance
Ingram Content Group UK Ltd.
Pitfield, Milton Keynes, MK11 3LW, UK
UKHW042144280225
455719UK00001B/86